SpringerBriefs in Molecular Science

Biometals

Series Editor

Larry L. Barton

For further volumes:
http://www.springer.com/series/10046

B. Rowe Byers
Editor

Iron Acquisition by the Genus *Mycobacterium*

History, Mechanisms, Role of Siderocalin,
Anti-Tuberculosis Drug Development

 Springer

Editor
B. Rowe Byers
Department of Microbiology
University of Mississippi Medical Center
Madison, MS
USA

ISSN 2212-9901
ISBN 978-3-319-00302-3   ISBN 978-3-319-00303-0   (eBook)
DOI 10.1007/978-3-319-00303-0
Springer Cham Heidelberg New York Dordrecht London

Library of Congress Control Number: 2013938745

Printed on acid-free paper

Springer is part of Springer Science+Business Media (www.springer.com)

Contents

Chapter 1
Introduction to the Book

B. Rowe Byers

1.1 The Value, Danger, and Capture of Iron by Bacteria

Astronomers tell us that all stars are furnaces fueled by fusion of hydrogen into helium. When their fuel is spent, the central cores of stars collapse, reaching 100 million degrees Kelvin. Under these conditions nuclei fuse to form heavier elements, such as carbon, oxygen, and eventually iron. Some of these dying stars then explode, scattering the heavy elements through space; subsequently, the remnants of the explosion re-condense to form new stars surrounded by planets composed of the heavier elements. Utilizing the electromagnetic force of nature (that is largely responsible for associations of atoms and molecules) and the lucky coincidence of the orbit of Earth and the mass of the sun, carbon bound with itself to yield our biological system. Other heavier elements were incorporated as integral members of the system; their capacities to catalyze chemical reactions and to bond to specific atoms were used to build a coherent entity of life. Iron became critical for metabolism.

Iron is valuable because it has two stable valence states that impart a wide range to the metal's reactivity. However, iron is hazardous. In a biological cell, free iron can catalyze the formation of hydroxyl radicals and peroxide, both of which can destroy biological molecules. All cells that utilize iron (and almost all do) must carefully handle internal iron to prevent escape of the metal. Moreover, acquisition of iron can be difficult. In an oxidized environment near pH 7 the oxidized ferric form precipitates from solution, a troublesome property requiring an efficient iron-capture mechanism. In microorganisms the most studied iron-gathering systems are those based on the production of low molecular mass ferric-chelating molecules called siderophores that bind and deliver iron to the cells. Like microorganisms, human systems work within the constraints of iron's solubility and reactivity and iron is present in almost a closed, protected circuit. This normal

B. R. Byers (✉)
University of Mississippi Medical Center, Jackson, MS, USA
e-mail: rowebyers@comcast.net

B. R. Byers (ed.), *Iron Acquisition by the Genus Mycobacterium*,
SpringerBriefs in Biometals, DOI: 10.1007/978-3-319-00303-0_1,
© The Author(s) 2013

human iron management process secludes iron from an invading pathogen, creating an iron-poor situation for the pathogen. Some pathogens use their siderophores to interrupt their host's iron circuit.

1.2 Scope of the Book

Our subject is the mechanism of iron acquisition by the genus *Mycobacterium*, a ubiquitous group that is composed of both saprophytic and pathogenic types, including the devastating pathogen *M. tuberculosis*. Almost all mycobacteria produce both a cell associated water-insoluble siderophore mycobactin and a modified mycobactin termed carboxymycobactin that is water soluble and excreted from the cells. The saprophytic mycobacteria elaborate and excrete yet another siderophore exochelin not found in the pathogenic types.

Siderophore discovery and isolation in the mycobacteria have an interesting history and Colin Ratledge, the authority on this subject, first will trace the historical aspects of siderophores in the mycobacteria from a unique personal viewpoint. His handiwork and that of other pioneer scientists interested in mycobacterial iron uptake launched the current studies, forming the platform on which these are based. It is unlikely that a unique work similar to Chap. 2 on the history of mycobacterial iron acquisition is, or will become, available.

Mycobacterial iron uptake is complex and excellent recent reviews on *M. tuberculosis* iron acquisition systems are available. In Chap. 3, I will attempt to summarize recent research on iron uptake in both pathogenic and saprophytic mycobacteria with special interest on the genetics of these processes.

Not only does the human iron handling system tend to withhold iron from microbial pathogens, other iron related host defenses are quickly mounted. Host production of siderocalin has been added as another player because siderocalin can bind and inactivate some siderophores, requiring a pathogen to either modify its siderophore or produce another siderophore that resists siderocalin inactivation. In Chap. 4, Benjamin Allred, Allyson Sia and Kenneth Raymond will detail the potential role of siderocalin in a mycobacterial infection.

The rise of multiple drug resistant strains of *M. tuberculosis* brings us to the crucial question for humanity: because iron is critical for mycobacterial growth, can what we know of the mycobacterial iron acquisition mechanisms be used to develop effective treatments for tuberculosis? Approaches to this question include synthesis of siderophore analogs and conjugation of siderophores with toxic agents. In Chap. 5, Raul Jaurez-Hernandez, Helen Zhu and Marvin Miller will describe the important recent developments in the use of the mycobacterial iron uptake processes in the potential development of anti-tuberculosis drugs.

Chapter 2
A History of Iron Metabolism in the Mycobacteria

Colin Ratledge

"Mycobacteria are nothing more than E. coli wrapped up in a fur coat"—Frank Winder.

Abstract The importance of iron to the metabolism of the mycobacteria was gradually appreciated during the first half of the last century. Frank Winder working in Dublin in the 1950s and 1960s was the first to establish the absolute amounts of iron needed for growth and, from his work, it was then possible to investigate the consequences of iron deficiency and subsequently how iron was solubilized and transported into mycobacterial cells. Parallel with this work, was the discovery by Alan Snow, at ICI Ltd, UK, of the mycobactins. These are essential growth factors for *Mycobacterium paratuberculosis* and their role in iron binding was then pivotal to elucidating the main aspects of iron uptake. However, mycobactins, being wholly intracellular materials, were unable to act as external siderophores for the solubilization of iron; this role was then found to be carried out by the exochelins discovered by the author of this review. The exochelins were of two types: those from the non-pathogenic mycobacteria were water-soluble pentapeptides whereas those from pathogenic species were modifications of mycobactin and were then named as the carboxymycobactins. The interdependency of these materials and others is then unraveled in this review. The review focuses mainly on the research work carried out over the last century leaving the present work on iron uptake to be covered in other reviews in this monograph.

Keywords Carboxymycobactin • Exochelin • Iron acquisition • Mycobacterial siderophores • Mycobactin

C. Ratledge (✉)
Department of Biological Sciences, University of Hull, Hull HU6 7RX, UK
e-mail: c.ratledge@hull.ac.uk

B. R. Byers (ed.), *Iron Acquisition by the Genus Mycobacterium*,
SpringerBriefs in Biometals, DOI: 10.1007/978-3-319-00303-0_2,
© The Author(s) 2013

3

2.1 Introduction

In writing this review I have attempted to cover as much as possible of the early research work of how our understanding of iron metabolism has developed over the past six or seven decades as this material now receives little, if any, mention in current research papers or reviews. However, there are many nuggets of valuable information tucked away in many of these early papers that should not be forgotten as they can help to identify aspects of the topic that still need investigation and explanation.

Iron metabolism does not, though, have a long history. The real beginnings only start in the 1950s through the work of Frank Winder. He was the pioneer in this area as he was the first to appreciate that, in order to study how iron was metabolized, it was necessary to know how much iron was needed by the mycobacteria to grow. It was then essential to prepare culture medium with as much iron as possible being removed from it; this then provided the first culture media that were genuinely, and knowingly, iron deficient. Thus, Winder's research group in Dublin, Ireland, was then the first group to be engaged on understanding the consequences of iron deprivation in mycobacteria. This deprivation of iron in the growth medium was the essential first step in understanding how iron was assimilated into the bacteria. Now, many laboratories throughout the world are actively pursuing a variety of aspects of the subject but these still have a long way to go before the problems of iron metabolism, both in vitro and in vivo, could be considered as being solved.

The importance of iron in the growth and multiplication of the tubercle bacillus, and other pathogenic mycobacteria has gradually been appreciated over the past three or four decades as it has been increasingly realized that iron-deficient growth conditions are the 'natural' state for pathogenic bacteria to be in when they are causing infections in a host animal. Thus, from the early work of Frank Winder from the 1950s and into the 1970s has developed a crucial understanding of how mycobacteria are able to acquire the iron that is essential for their growth and multiplication when causing tuberculosis and related mycobacterial diseases.

The other pioneer in this field was Alan Snow who was involved in the discovery of the mycobactins that were subsequently to become central to the iron metabolism story. However, the involvement of the mycobactins with iron only came in the 1960s, some 20 years after the initial descriptions of the molecule. By this time, interest in mycobactins, as possible target molecules for the design of rational anti-tuberculosis agents, had faded and with only one or two groups then pursing the biochemical puzzles that had been opened up by the discovery of these unique microbial siderophores. Interestingly, however, these early thoughts of synthesizing novel mycobactin antagonists are now being re-visited as the need for novel chemotherapeutic agents for the treatment of tuberculosis becomes ever-more urgent.

I hope, therefore, that this early history of the subject will be of interest to readers. I have placed much less emphasis on developments over the past 10–15 years as, although a lot is now happening in many laboratories, these aspects are mainly focusing on filling in the details of individual parts of the overall picture and are, in any case, covered elsewhere in this monograph. Solving the problems of iron metabolism is a little like doing a jigsaw puzzle: in this case, however, one is never sure that one has all the pieces available to provide the final picture. One tries to provide a coherent and comprehensible view of iron metabolism in the mycobacteria from the pieces that can be found. But how many more are still missing? In spite of us thinking that we now know what the final picture is going to look like, there is clearly still a long way to go!

2.2 The Trouble with Iron

Iron is an essential trace element for almost all living cells. The only exceptions appear to be some lactobacilli and the spirochete, *Borrelia burgdorferi,* that is the causative agent of Lyme disease. Iron is needed as an essential co-factor in many enzymes and is also the critical metal ion in all haem compounds, including all the cytochromes that carry out essential functions in energy metabolism and also are components of several key enzymes. Iron, however, is unique amongst the nutrients needed for cell growth in that is insoluble at neutral pH values. However, this needs to be qualified as iron exists in two states: the reduced ferrous form and the oxidized ferric form. It is the latter form that is insoluble and, although ferrous salts are water-soluble, they quickly oxidize to the ferric form, a reaction which is accelerated if the iron is in a chelated form. Although the solubility of ferric iron at pH 7 has usually been stated to be 10^{-18} M, more recent measurements give this as about 10^{-9} to 10^{-10} M [1, 2]. This revised lower value arises because it is now appreciated that the principal ionic species that exists in aqueous solution is $Fe(OH)_2^+$ and not $Fe(OH)_3$ as previously thought. Even though this revised value is a billion times higher than the earlier value, it still results in iron being effectively insoluble as 10^{-9} M corresponds to 56 pg/ml. This then effectively renders iron as being unavailable to cells. Specific mechanisms have therefore evolved so that cells may acquire iron from the environment and also hold it within themselves in a usable form. These mechanisms differ between animals, plants and microorganisms.

For microbial pathogens, solving the problem of iron acquisition is essential. If they cannot acquire iron from the sources of iron inside the host that they have infected, then they will be unable to grow and thus cause disease. As pathogens do cause disease, we can obviously conclude that all pathogens must have evolved mechanisms for iron acquisition. The principal sources of iron within an animal are: transferrin, ferritin, haemoglobin and haem-containing proteins.

Transferrin is the principle iron transporting protein in the blood. There are related proteins of lactoferrin, found in milk and other extracellular fluids and secretions, and ovaferrin (formerly known as conalbumin) that is found in eggs. These are large proteins (~80 kDa) but only have two binding sites for ferric iron.

Ferritin is a protein (~50 kDa) comprising 24 identical subunits that form a hollow sphere into which up to 4,000 atoms of Fe(III) can be stored. This is the principal form of iron storage in animal cells. A similar form of it, bacterioferritin, exists in bacteria, including mycobacteria, for the same function.

In haemoglobin and other haem containing molecules iron is tightly bound as the ferrous ion and therefore, for metal release, the molecule must be degraded. This occurs principally with haemolytic bacteria which therefore excludes most pathogenic mycobacteria.

To achieve release of iron from transferrin and ferritin requires that pathogenic bacteria, including mycobacteria, must either attack the protein itself by secreting various proteases, use a ferric reductase that would generate ferrous ions that might then be directly assimilated, or, alternatively, use molecules of very high iron binding strength that can then, literally, strip the iron out of the molecules. The materials that can do this are known as siderophores and those relevant to the mycobacteria will be covered in this review and also elsewhere in this monograph.

Thus, we can conclude that all mycobacteria, whether pathogenic or saprophytic, require iron and that the acquisition of iron from whatever source requires specific mechanisms for achieving this.

2.3 Iron as an Essential Nutrient for Mycobacteria

In the early days of mycobacteriology, iron was 'guessed' to be an essential minor trace element and workers, such as Sauton [3], in devising appropriate growth medium for the cultivation of mycobacteria, recommended that iron be added to the culture medium at 10 μg/ml. Obviously, this was an empirical amount but later workers [4–6] were able to confirm that iron was indeed required to achieve full growth of various mycobacteria. Edson and Hunter [4] indicated that, for *M. phlei*, 3.75 μg Fe/ml was needed for full growth but Turian [6] revised this to just 1 μg/ml. Clearly, the amount of iron added to the medium would depend on the amounts of iron which were adventitiously included by other ingredients of the medium. Nor should the presence of iron in the water or in the glassware being used for cultivation be ignored. Thus, to establish what amounts of iron might be necessary for growth it was first necessary to prepare culture medium that with as little iron as possible. This was first appreciated by Frank Winder who then pioneered a series of in-depth studies on the role of iron in the metabolism of the mycobacteria.

Frank Winder (Fig. 2.1) carried out all his work on the mycobacteria at Trinity College, Dublin, Republic of Ireland, starting from about 1953 and continuing to his death in 2007 at the age of 79. Initially, Winder's work was done in the Laboratories of the Medical Research Council of Ireland and then later in the Department of Biochemistry also within the College. In a seminal paper in which the conditions of iron deficient (and also zinc deficient) growth of a mycobacterium (*Mycobacterium smegmatis*) were first described, Winder and Denneny [7] observed that when some, but not all, cultures of the bacterium were grown in modified Proskauer and Beck medium to which no iron or other trace elements

Fig. 2.1 Frank Gerald
Augustine Winder (1928–
2007)

had been added (simply because the formulation of this medium did not include addition of an iron or zinc salt), the cells prematurely ceased growth, failed to form the usual pellicle and became elongated with a low level of DNA. The cells had, in fact, been cultivated in a medium that was accidentally deficient in both iron and zinc. In addition, the tubes being used for the cultivations had been recycled a number of times thereby exhausting adventitious metal ions from the glassware itself. When both iron and zinc were subsequently included in the medium, full and normal growth of *M. smegmatis* was restored including restoration of DNA synthesis.

This discovery then opened up the door to a major study of iron metabolism in the mycobacteria by Winder and his associates that also included the author of this review (see Fig. 2.2) who joined his team as a post-doctoral fellow in 1960 and worked in the MRC of Ireland Laboratories until 1964. Winder and O'Hara [8, 9] then described the effects of both iron and zinc deficiencies on the composition of *M. smegmatis*; some of the key findings are summarized in Table 2.1. However, before these studies could begin, it was necessary to devise a simple protocol for the removal of the trace metal ions from the medium otherwise there would be an

Fig. 2.2 Colin Ratledge
(1936–)

Table 2.1 Metabolic consequences of iron deficiency in mycobacteria

Consequences of deficiency	References
Long forms of cells produced; DNA synthesis declines	Winder and O'Hara [8]
Increased activities of DNA repairing enzymes	Winder and Coughlan [12, 13]
	Winder and McNulty [14]
	Winder and Barber [17]
Activity decrease of iron-containing enzymes	Winder and O'Hara [9]
Low content of cytochromes *a* and *b* with low content of coproporphyrin	McCready and Ratledge [20]
Increased biosynthesis of iron chelating compounds.	
• salicylic acid	Ratledge and Winder [50]
• mycobactins	Snow [26]
• exochelins and carboxymycobactins	Macham and Ratledge [88]
	Macham et al. [88, 89]

inadequate basis on which to conduct the subsequent studies. The procedure used, which was derived from the work of Donald et al. [10], was to autoclave 2 L of medium in a 5 L flask with 1 % (w/v) high-grade, activated alumina. When the flask came out of the autoclave, the contents was immediately shaken thoroughly and then, when cool, filtered through Whatman number 542 filter paper which was considered to be the highest quality, ash-free paper then available. The first 50 ml of the filtered medium was discarded as this has come into contact with measuring cylinder, the filter funnel and the filter paper itself. The remaining medium was then dispensed in 100 ml lots into cleaned 250 ml conical flasks. Cleaning the flasks necessitated devising another strategy. Initially flasks were cleaned with chromic acid although this procedure was soon replaced by filling the flasks with alcoholic KOH and leaving overnight. This was followed by washing them in distilled water then standing for another night in 2 M HNO_3. The flasks were finally thoroughly rinsed in distilled water and allowed to drain upside down before being filled with medium the next day. The procedure was extremely tedious and time-consuming and services of a dedicated technician were then needed to do all the preparatory work.

Winder and O'Hara [11] also carried out some significant analytical work on the cellular content of iron in the mycobacterial cells. Under the most stringent iron deficient growth conditions, *M. smegmatis* contained 64 μg Fe/g cell dry weight suggesting that this was the lowest possible concentration needed for the cells to function. Obviously, as the mycobacteria have an essential requirement for iron, the cells would not be able to grow if there was, literally, no iron in the medium. They could not synthesize the various cytochromes and iron-containing enzymes that are vital for cell metabolism and growth. If iron was not limiting, then the iron content of the cells rose to 224 μg/g cell dry wt. Values for the zinc content of the cells were simultaneously calculated as 11 and 43 μg/g cell dry weight, respectively. It was evident, however, that iron-deficiently growing cells were adapting to allow some growth to occur but clearly major changes were occurring within the metabolic pathways to minimize the detrimental effects of iron deficiency.

One of the main effects of iron deficiency on metabolism appeared to be a decrease in the DNA to protein ratio [8]. This was subsequently attributed to there being a considerable increase in the activity of an ATP-dependent DNAase [12, 13] and a DNA polymerase [14]. Further work on the DNAase [15, 16] considered that it was involved in recombination repair and possibly in excision repair. Interestingly, how the increased activities of both these enzymes then correlated with the original discovery, that there were low concentrations of DNA in iron-deficient cells, was resolved by Winder and Barber [17] who reported that hydroxyurea could induce the same effects as iron deficiency, including cell elongation. However, one cause of DNA degradation might be in the lowered activity of ribonucleotide reductase which, in *E. coli*, is known to contain iron as an essential co-factor [18] and would therefore lead to an alteration in the pool of nucleotides. This possibility, though, does not appear to have been followed up in *M. smegmatis* and the general consensus was that the decrease in DNA during iron deficient growth was a secondary but not a primary effect.

The paper by Winder and Barber [17] and a subsequent one by MacNaughton and Winder [19] were the last papers that Frank would write on aspects of his work connected to iron deficiency in the mycobacteria. He then, with respect to his continuing interest in the mycobacteria, concentrated on trying to unravel the mechanism of action of isoniazid (INH) as one of the more potent anti-TB compounds then available. This work had also begun in the 1960s and was carried out in parallel with the iron deficiency studies.

It was not, though, surprising that Winder and O'Hara [9] reported considerable decreases in activity of many iron-containing enzymes. These included various cytochromes. These findings were subsequently confirmed by McCready and Ratledge [20] also working with *M. smegmatis*. Non-haem iron in the cells dropped to ~0.2 nmol/g CDW from ~5 nmol/g CDW and cytochromes *a* and *b* could not be detected. However, cytochrome *c* was scarcely affected, as were the flavoproteins, indicating that these iron-containing components were, to some extent, protected from the severest ravages of iron deficiency. Presumably these components have very high affinities for iron and thus can acquire iron even when it was available in the smallest concentrations inside the cell and against competition from other iron-dependent cytochromes and enzymes.

Iron deficiency therefore produces a major diminution of most components in the respiratory chain and this, in turn, will inevitably cause a decline in activity of glycolytic enzymes as the final oxidation of pyruvate, via the tricarboxylic acid cycle and its linkage to oxidative phosphorylation that involves many cytochromes, would be seriously impaired by iron deficiency. Thus, it is not unexpected to find a general down-regulation of many enzyme activities in the central pathways of metabolism simply as a consequence of there being diminished energy (ATP) production.

It was apparent from these experiments of the 1960s and 1970s that iron deficiency in mycobacteria was causing the cells to become 'anaemic', somewhat equivalent to the condition that seen in humans and other animals. Cells became 'lethargic': they had a diminished supply of energy, failed to grow properly and failed to carry out normal metabolism. They also become noticeably much paler than cells grown with a surfeit of iron. McCready and Ratledge [20] and McCready [21] found that the content of porphyrins in iron deficient cells was adversely affected: in *M smegmatis*, coproporphyrin III was less than 25 μmol/g CDW after iron deficient growth compared to over 200 μmol/g CDW in cells grown iron sufficiently. It is then this absence of porphyrin that accounts for the very pale appearance (the 'anaemic' condition) of the iron-deficient cells. This low content of porphyrin was later also observed in iron-deficiently grown *M. avium* [Barclay and Ratledge, unpublished work in the 1980s] and may then be a general explanation for most mycobacteria being much paler when grown without adequate amounts of iron. This phenomenon, which is not seen with other bacteria, must be caused by repression of the biosynthesis of the porphyrin nucleus due to lack of iron in the cells. This makes metabolic sense. Why synthesize something that cannot be converted into the end-product: haem? Therefore stopping the synthesising of the precursor of haem is a sensible metabolic strategy under iron deficient conditions.

But this then poses a major problem to the cells: if iron then becomes available to the cells for whatever reason, what are the cells going to do with this iron if it cannot be immediately converted into haem because there are no precursor molecules of porphyrin available? Up-regulation of porphyrin biosynthesis cannot be immediate. The cells must therefore have some means of acquiring the iron and holding it in a form which can then be mobilized as porphyrins begin to be re-synthesized. This aspect of iron metabolism then is considered in the next section of this review.

Such is the need to scavenge whatever iron might be available from the environment, so as to rescue the cells from their 'anaemia', that it is therefore not surprising to observe that much metabolic effort is expended by the iron-deficient mycobacterium to acquire iron from its environment. Iron deficiency, however, is not just a man-made artificial construct for laboratory cultivation experiments. There is good reason now to consider that iron deficiency is the normal status of pathogenic mycobacterium within the animal tissue which it is infecting. Pathogenic mycobacteria therefore must overcome the natural defenses of the infected host animal that seeks to withhold iron from the invading bacteria. Unless a pathogen, and not just a mycobacterium, can acquire iron from its host, it will not be able to grow and thus become pathogenic. Gaining iron is therefore possibly the first step for an invading bacterium to achieve in order to grow in vivo. Thus, the mechanisms of iron acquisition by mycobacteria are of prime concern if we are to understand anything about the pathogenicity of these bacilli.

2.4   Early Discoveries of the Major Components of Iron Acquisition by Mycobacteria

2.4.1   The Mycobactins

The first clues about how iron might assimilated by mycobacteria came very indirectly from the initial observations by Twort and Ingram [22–24] when they were attempting to cultivate the mycobacterium that was the causative agent of Johne's disease in cattle. However, it would be more than 50 years before it would be appreciated how these early observations, and the subsequent discoveries arising from them, fitted in with iron metabolism to unravel a major and unique feature of mycobacterial metabolism.

Johne's disease in cattle causes chronic enteritis and was found to be caused by a mycobacterium that was then called *M. johnei* [25] but was re-named as *M. paratuberculosis*. This name is therefore used in the remainder of this review although more recent taxonomic work has re-classified the bacillus yet again, as mentioned below. The organism, very importantly for the iron assimilation story, could not be cultivated in ordinary laboratory medium but Twort and Ingram [22–24] found that by supplementing the egg-based medium they were using with dry, killed human tubercle bacilli they could then achieve good growth of this previously uncultivatable *Mycobacterium* species. Animal tissues and extracts were ineffective. They subsequently found that other killed mycobacteria could also support growth: these included *M. phlei*, *M. smegmatis*, and *M. butryicum* as well as other less-well defined mycobacteria. Also extracts from the killed mycobacteria prepared using organic solvents were equally successful in promoting growth. The conclusion was reached that *M. paratuberculosis* lacked the ability to synthesize some essential growth factor but that this material was synthesized by several competent mycobacteria that could be grown in laboratory culture medium. As was subsequently acknowledged by Snow [26], Twort could be considered to be "... a true pioneer in this field as the concept of vitamins and growth factors were only dimly recognized at that time". Yet here was a scientist proposing a radical concept of a transferable growth factor, available from some microbial sources but not synthesized by the host microorganism. This was exactly the same concept that lay behind the discovery of many of the vitamins essential for human metabolism but this was a novel concept for achieving growth of a microorganism.

Twort's findings, however, were not developed for another 30 years. In the meanwhile, cultivation of *M. paratuberculosis*, which was, and continues to be, a considerable veterinary problem, in laboratory media was routinely achieved by adding a simple extract prepared from cells of *M. phlei* which, of course, is a non-pathogenic species. Cultivation of Johne's bacillus did not, therefore, require a growth factor that was confined to pathogenic mycobacteria. A saprophytic mycobacteria could do just as well and was obviously less hazardous to grow and extract.

We now must move to the 1940s for the development of Twort's observations and to the laboratories of one of the major industrial chemical companies in the UK: Imperial Chemical Industries (ICI) Ltd at their pharmaceutical research laboratories at Blackley near Manchester and later at Wilmslow and finally at Alderley Edge in Cheshire. An account of the thinking that went on in ICI for Twort's discoveries to become a priority research program has been given by Snow [26]. The person who initially took up the baton was J. Francis who pointed out in 1945, (cited by Snow [26]) that a specific growth factor for *M. paratuberculosis* was being synthesized by *M. tuberculosis*. As no other microorganism, other than another species of mycobacteria, could produce such a compound, then this compound must be unique to the mycobacteria. In addition, as Twort had found that extracts of animal tissues, including those from cattle, could not support growth of Johne's bacillus, this growth factor was not being synthesized by animals.

As there was no chemotherapeutic treatment for tuberculosis in 1945, Francis's reasoning was clear: *M. tuberculosis* and other mycobacteria were synthesizing a specific growth factor not found in humans. If this compound could be identified then it opened up the opportunity of designing an appropriate antagonist that would then, hopefully, be specifically inhibitory to the tubercle bacillus. Such inhibitors should not though affect the infected human as there was no suggestion that the missing growth factor for *M. paratuberculosis* was synthesized in animals. ICI Ltd obviously considered that if this aspiration could be realized then it could be an extremely lucrative project and it is little wonder that major efforts were expended to attain the goals.

GA Snow (see Fig. 2.3), always known as Alan, joined the team in the late 1940s and was to become the major driver of the entire project. He has written that the initial work on isolating and purifying the growth factor was undertaken by J. Madinaveitia and H.M. Macturk working with the initial project leader, J. Francis [27]. They used massive quantities of *M. phlei*: in all some 50–60 kg dry cells were used and the growth factor was eventually isolated and purified. It was given the name mycobactin.

Fig. 2.3 G Alan Snow (with kind permission of The Biochemical Society, UK)

The work, though, was extremely difficult. It was hindered by there being no adequate assay for the growth factor as its structure was unknown and, although it could bind to metal ions, this was not regarded of any particular significance. Indeed, in the first full-length paper describing the isolation of mycobactin by Francis et al. [28], more attention was given to the copper complex of mycobactin and this was, in fact, the complex of mycobactin that was studied in detail. Iron binding received only one mention in this seminal paper: "Ferric salts react with mycobactin to give an intense reddish purple colour". How these workers missed the significance of iron binding to mycobactin, or failed to appreciate that this would provide a simple assay for quantifying mycobactin, now seems strange because the preparations of unchelated mycobactin readily form the red ferric mycobactin complex, having a tenacity for iron binding is so strong that the mycobactin readily stripped iron from water, glassware and any materials containing trace amounts of iron. However, as Snow was using such large quantities of cells and extracting considerable amounts of mycobactin, the material was only turning a light brown in color as clearly the amount of available iron was relatively small.

The early work on mycobactin was helped considerably by mycobactin adventitiously forming a crystalline aluminium complex [27] during its purification and passage through a column of chromatographic alumina. This allowed the team to carry out some X-ray crystallography giving a suggested molecular weight of about 914. This was subsequently revised to 870 with the formula of $C_{47}H_{75}O_{10}N_5$. To work out the structure of such a large molecule, however, was then a considerable and daunting task. Also, as mentioned above, Snow and his colleagues had found that it was extremely difficult to assay mycobactin during its purification processes. Initially, assays for its presence had to rely on its growth-stimulating properties for *M. paratuberculosis* which were slow and extremely tedious even with improved techniques [29]. There were, however, suggestions by Antoine et al. [30] and Reich and Hanks [31] that *Arthrobacter terregens*, which also required a 'terregens factor' for growth, might be a more suitable organism for the bioassay of mycobactin as it could be grown in about 3 days or so. This bacterium was, though, much less sensitive to mycobactin than *M. paratuberculosis* [26] and was also responsive to growth factors other than mycobactin. It therefore does not appear to have been used to any great extent by Snow himself.

The initial major effort that had been put into the pharmacological aspects of the project appears to have dissipated somewhat by the very early 1950s and the original authors of the 1953 paper [28] do not appear, even in the acknowledgements, of the next papers that were published on the structure of mycobactin in 1954 [32, 33]. Perhaps the discovery of streptomycin in 1943 as the first anti-tuberculosis antibiotic followed by its general availability in the late 1940s, together with the arrival of PAS (*p*-aminosalicyclic acid) as a second anti-TB agent in 1953, may have influenced the senior managers of ICI that the future of anti-tuberculosis treatment would lie with antibiotics and not with problematic, and still to be synthesized, possible antagonists of a still largely uncharacterized mycobactin. But for whatever reason, Alan Snow then was the person who almost single-handedly elucidated the structures not only of the mycobactin from *M. phlei* but many other ones as well (Fig. 2.4).

The initial description of a possible structure for mycobactin form *M. phlei* was given in the December 1954 issue of the Journal of the Chemical Society [33]. However, there was a problem with working out how one of the hydrolytic products of mycobactin, 2-amino-6-hydroxyaminohexanoic acid, was orientated in the molecule. Two possible structures for it were offered. It was though another 11 years before this issue was resolved. No full-length papers were published by Snow or any other person on the mycobactins from 1954 to 1965 though there was a short preliminary communication made in 1961 to a meeting of the Biochemical Society in the UK [34] concerning the isolation of the mycobactin from *M. tuberculosis*. But it seems likely that the mycobactin project was now de-prioritized roundabout this time and, in a letter written to Philip D'Arcy Hart at the Medical Research Council Laboratories in London and dated July 29th 1968, Snow himself said that "We have revived some interest in this topic after a lapse of a number of years". This would suggest that the project had been completely abandoned in the late 1950s and early 1960s. Sometime then in the 1960s, interest in mycobactins must have then re-started but, in all probability, only Alan Snow, with possibly just one or two technical assistants, would have been engaged on the project for most of the remainder of the program.

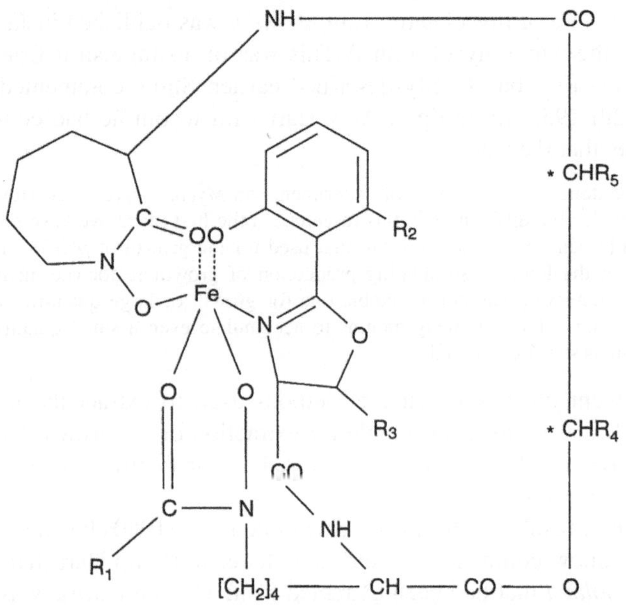

		Substituents				
Organism	Mycobactin	R_1	R_2	R_3	R_4	R_5
M. aurum	A	13Δ	CH_3	H	CH_3	H
M. fortuitum	F†	17,11Δ	H	CH_3	CH_3	H
M. fortuitum	H†	19,17Δ	CH_3	CH_3	CH_3	H
M. marinum	M†	1	H	CH_3	$C_{17}H_{35}$	CH_3
M. marinum	N†	2	H	CH_3	$C_{17}H_{35}$	CH_3
M. phlei	P	17cisΔ	CH_3	H	C_2H_5	CH_3
M. terrae	R	19Δ	H	H	C_2H_5	CH_3
M. smegmatis	S	17,15cisΔ	H	H	CH_3	H
M. tuberculosis	T	19Δ	H	H	CH_3	H
M. avium	Av	Δ2 alkenyl	H	H	$C_{10}H_{23}$?	CH_3
M. intracellulare	Av‡	Δ2 alkenyl	H	CH_3	satd alkyl	CH_3
M. scrofulaceum	Av‡	alkenyl	H	H	satd alkyl	CH_3
M. paratuberculosis	Av‡	Δ3 alkenyl	H	CH_3	satd alkyl	CH_3
M. paratuberculosis	J	15Δ	H	H	isopropyl	CH_3

Fig. 2.4 General structures of the ferric-mycobactins [26, 64, 67]. The substituents at R1 are usually alkenyl chains with a *cis*-double bond at C2 (exceptions are for mycobactins M and N, both from *M. marinum*). There are usually a number of chain lengths, only the major ones are given; † indicates two distinct mycobactins from the same strain, ‡ mycobactins are considered as being equivalent

The structure of mycobactin from *M. phlei* was published in January 1965 [35] and given the name mycobactin P. This was to distinguish it from mycobactin T which Snow [34] had briefly described earlier. Snow commented in letter dated August 12th 1955, to Philip D'Arcy Hart with whom he had considerable correspondence, that the team

"... have done quite a number of experiments on *Myco. tuberculosis*. However, we have met considerable difficulties in this research. In the first place, we have grown our tubercle bacillus on a medium similar to that used for the growth of *phlei* with beef infusion present in the hope of stimulating production of growth factor (i.e. mycobactin). This medium, however, was not a suitable one for growth of large quantities of the tubercle bacillus, and it took us many months to accumulate even a small quantity of the dried organism as starting material."

Snow went on to say that the methods used to extract the mycobactin from *M. phlei* "... were quite unsuitable for extraction of the growth factor from tubercle. We have also been severely hampered in our facilities for testing the activity of the concentrates."

The initial results of the work being done in the 1950s had also produced some problems. Snow commented in the same letter to Philip Hart that the growth factor for *M. johnei* that had been extracted from *M. tuberculosis* was similar but not identical to the mycobactin from *M. phlei*. Thus, the magnitude of this project that was facing Alan Snow cannot be over-stated.

The structure of mycobactin T itself was eventually solved and published in 1965 by Snow [36]. Hough and Rogers [37] were subsequently able to confirm in detail Snow's structure and stereochemistry of mycobactin P by using X-ray crystallography. The ferric ion was found to lie in a V-shaped cleft with a very strained octahedral configuration involving five oxygens and one nitrogen. The exceptional stability of mycobactin with ferric iron was then explained together with explaining how the iron could be easily released from mycobactin by reduction of ferric to ferrous iron [38–40] where the resultant ferrous ion had little or no affinity to mycobactin and would thus be available for incorporation in apoenzymes and other proteins.

It was only in the two major papers of 1965 by Snow that dealt with the structures of mycobactins [35, 36] was iron binding recognized as a major attribute of them. Snow [35] now appreciated that mycobactin was, in fact, a microbial siderophore—or what were then called 'sideramines'. He commented that the isolation of the desferri-form was directly attributable to the cultivation medium being used, that was beef infusion broth, having a low content of ionized iron. In other words, and with hindsight, the iron-containing components of the medium would be various haem compounds and also ferritin and transferrin that would withhold iron from bacteria and, therefore, the cells were accidentally being grown iron deficiently. This was confirmed by Norman Morrison [41, 42], at the Johns Hopkins-Leonard Wood Memorial Laboratory, Baltimore, USA, who in a personal communication to Alan Snow, reported that large amounts of mycobactin could be produced by growing *M. phlei* in a synthetic medium with less than 0.2 μg iron/ml. This then indicated a much easier way to optimize the accumulation of mycobactins in mycobacteria.

Following the descriptions of the structures of the first two mycobactins, P and T, [35, 36] the structures of other mycobactins were elucidated during the remaining years of the 1960s by Snow who had now been joined by a very able technical assistant, Mr. A. J. White. Mycobactins S and H types from, respectively, *M. smegmatis* and *M. thermoresistible* [43] were described; this was followed by descriptions of the mycobactins from *M. aurum*, *M. terrae*, *M. fortuitum* and *M. marinum* that were labeled as types A, R, F and M and N, respectively [44]. The structures of all the mycobactins that were determined by Alan Snow are given in Fig. 2.4. It is probably worth repeating the observations made in this final paper that the mycobactins M and N, both from *M. marinum*, were inhibitory towards *M. tuberculosis* and these two mycobactins were distinctively different from the mycobactins from other mycobacteria in having the characteristic long alkyl chain attached to a different part of the molecule (Fig. 2.4). This observation, however, does not appear to have been taken up by any other group looking to realize the potential of designing inhibitors of *M. tuberculosis* when this work of Snow and colleagues began 25 years earlier. These papers were, in fact, the last significant publications arising from the work at ICI Ltd. Alan Snow, himself, wrote a review of the mycobactins which remains the definitive account of the chemistry and major properties of these iron-binding compounds [26]. It is still quoted in many research papers today. This review contains many details that are still salient today; this includes considerable information on the binding of metal ions to the molecule including, of course, iron. The tenacity of mycobactin for iron is the major feature of the molecule and Snow, on the basis of desferrimycobactin being able to remove the iron from ferric-desferrioxamine B [45], calculated that its stability constant was well in excess of 10^{30}. The mycobactin project at ICI Ltd came to a close at the end of the 1960s with Snow himself then retiring about a decade later. The original objectives of the research, however, had not been fulfilled though considerable interest is still evident to-day in looking at aspects of iron metabolism in the mycobacteria for opportunities to design novel anti-tuberculosis agents. This topic is then re-visited by other contributors in this monograph.

Alan Snow appears to have received many requests for samples of mycobactin P once it was firmly established that this was the growth factor essential for the growth of johne's bacillus. I have already commented on some correspondence between Alan and Philip D'Arcy Hart at the MRC Laboratories in London. Hart was not only wanting to grow *M. johnei,* as it was still referred to then, but also the leprosy bacillus, *M. leprae*, and the rat leprosy bacillus, *M. lepraemurium*. Also pursuing the same objectives was John Hanks at Johns Hopkins University, Baltimore, who also had received samples of mycobactin from Snow. Unfortunately, in spite of many attempts both in London and Baltimore, the leprosy bacillus in neither location showed any sign of growth in mycobactin-supplemented medium.

The correspondence between Hart and Snow began in 1955 and continued up to 1972 with Snow finally concluding (October 26th, 1972) that "(o)ur own stocks of mycobactin P are quite low now, because we have given so much away but we can still help (you) with small quantities when required (in a good cause!)". The author of this chapter has been privileged to receive the original letters concerning the mycobactins that had been sent and received by Philip Hart. Hart published one paper dealing with the growth of *M. johnei* in a mycobactin-containing medium [46]. Philip died in 2006 aged 106 having remained active in research until his final 2 years.

The question of how *M. paratuberculosis* was able to grow in vivo and acquire iron from the host was not solved until much later although Norman Morrison, working at the Johns Hopkins University, found that *M. paratuberculosis* could, in fact, grow without mycobactin if the pH of the growth medium was dropped to 5.5, or even to 5.0 as was later reported by Lambrecht and Collins [47]. This, it was suggested, might then mimic the conditions of growth in vivo and might be enough to increase the solubility of free iron to the point where it could now be acquired without the need for mycobactin. A (partial) resolution of this problem came, however, by the demonstration by Barclay and Ratledge [48] that *M. paratuberculosis* and other mycobactin-dependent strains of *Mycobacterium* could synthesize the extracellular counterpart to mycobactin, that is carboxymycobactin, and this would then be the way in which iron was acquired by the cells.

2.4.2 Other Mycobactins and Related Lipid-Soluble Siderophores

Once Alan Snow had solved the structure of mycobactin and worked out the main ways in which it could be extracted and characterized, the way was open for other workers to build on this work and examine other mycobacteria and genera related to the *Mycobacterium* genus for the presence of other iron binding compounds.

The presence of materials similar to mycobactin was noted by Patel and Ratledge [49] in species of *Nocardia*. The genus of *Nocardia* is taxonomically closed related to the mycobacteria and both are members of the actinomyces group of bacteria and have similar cell walls with a high content of lipid. It is therefore not too surprising that similar iron-binding compounds would then be found in these species. These were then named nocobactins and the structure of that from *N. asteroides*, termed nocobactin NA, was then elucidated by Ratledge and Snow [50]. This is shown in Fig. 2.5 where it is labeled as type *a*. It resembled the structure of mycobactin M (Fig. 2.4) but had a distinctive oxazole ring instead of an oxazoline ring and with a shorter alkyl chain. This type of nocobactin was also detected in *N. paraffinae, N. sylvodorifera* and *N uniformis* [51]. Two other types of nocobactin were found: type *b* (which is then known as nocobactin NB) was from *N. brasiliensis*, and type *c*, or nocobactin NC, was from *N. caviae* and *N. phenotolerans*. The latter types were similar to mycobactins S and T but had saturated alkyl chains rather than the unsaturated chains of the latter materials.

Once it was appreciated that iron deficient growth of a mycobacterium would increase the content of mycobactin by up to a 100-fold, it was then a relative easy matter to develop simpler methods of producing it. In our own work with

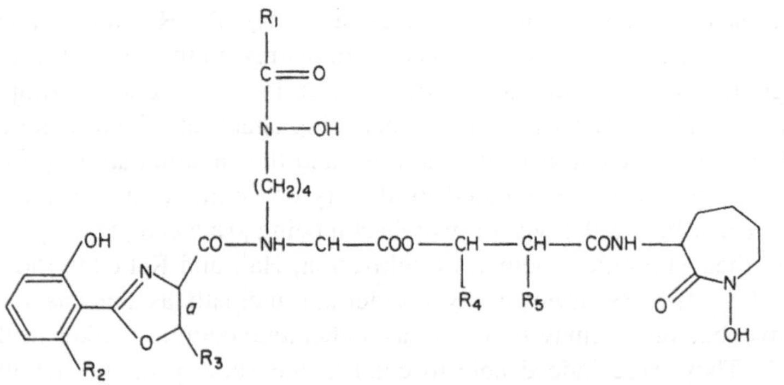

Fig. 2.5 Structures of nocobactins from *Nocardi* species [49, 51]. Type *a* from *N. asteroides*, *N. sylvoderifa*, *N. paraffinae* and *N. uniformis* (now all considered to be equivalent to *N. asteroides*) type *b* from *N. brasiliensis* and type *c* from *N. caviae* and *N. phenotolerans*. Types *b* and *c* are equivalent in structure to the mycobactins (see Fig. 2.4) but type *a* has its long alkyl chain at *R4* and is somewhat equivalent to mycobactins M and N (see Fig. 2.4) but also has an unusual oxazoline ring instead of the more usual oxazole ring

| | | | Substituents | | | |
	a	R_1	R_2	R_3	R_4	R_5
Nocobactin *a* type	Double bond	CH_3	H	CH_3	$C_{11}H_{23}$	CH_3
Nocobactin *b* type	Single bond	alkyl (saturated)	H	CH_3	CH_3	H
Nocobactin *c* type	Single bond	alkyl (saturated)	H	H	CH_3	H
Mycobactin P, S and T types	Single bond	alkyl (unsaturated)	H or CH_3	H or CH_3	CH_3 or C_2H_5	H or CH_3
Mycobactin M type	Single bond	CH_3	H	CH_3	$C_{17}H_{35}$	CH_3

M. smegmatis, it was possible to produce cells with 10 % (w/w) of mycobactin, a truly massive over-production of a siderophore. Richard Hall, one of my very able graduate students, was also able to achieve good yields of mycobactin by growing cells on agar plates (though highly purified agar had to be used) but there was no need to de-ferrinate the culture medium beforehand [52]. Cells could then be scraped off just one or two plates and then the mycobactin extracted with ethanol in the usual way. Up to 10 mg mycobactin per plate could be attained. It was then possible to analyze the mycobactins very quickly and easily using both thin layer chromatography and the newly arrived technique of high-pressure chromatography (HPLC) [53]. The former technique, with appropriate solvents, separated the mycobactins according to differences in the R_2, R_3, R_4 and R_5 nuclear sub-stituents (Fig. 2.4) and HPLC separated them mainly on the basis of differences in the length of the alkyl chain at R1. Bosne et al. [54] subsequently simplified this procedure by adding EDDA as an iron chelating agent into liquid culture medium to produce iron-deficient growth conditions and thus a stimulation of mycobactin production. They used this method to identify 65 strains of *M. fortuitum* and *M. chelonae* according to the type of mycobactin being produced [55].

Using their simplified method of cultivation, Hall and Ratledge [56] analyzed the mycobactins of 39 strains of mycobacteria, principally as a means to determine if the mycobactins "...may be useful as a chemotaxonomic marker in the myco-bacteria". They were indeed able to confirm the validity of the hypothesis and found that the mycobactins were strongly conserved molecules showing strong intra-species consistency and could be therefore used as chemotaxonomic charac-ters of high discriminatory power. The structures of the new mycobactins were not determined, however. Using the same method, Leite et al. [57] were able to iden-tify various clinical mycobacterial isolates from their mycobactins and Barclay et al. [58] were able to suggest an even more rapid method of detecting and identify-ing mycobacteria using [55]Fe-labelling of the mycobactins. A list of species pro-ducing mycobactin is given in Table 2.2; it is probably reasonable to conclude that most species of mycobacteria will be found to produce a mycobactin but there are some exceptions.

Rich Hall went on to apply his techniques to other groups of mycobacteria, showing equivalence of the mycobactins from *M. senegalense, M. farcinogenes* and *M. fortuitum* but a distinction from that from *Nocardia farcinica* [59]. He also examined the mycobactins from seven strains of armadillo-derived mycobacte-ria (ADM) [59] that were of interest because of the association of *M. leprae* with the armadillo and because, but unlike *M. leprae*, these bacteria could be grown in laboratory medium. The ADM were found to be a heterogeneous group; four of them produced materials that resembled the mycobactins from *M. avium-intracel-lulare-scrofulaceum* (MIAS) complex of mycobacteria suggesting that they could be assigned to this taxonomic grouping.

The MIAS group of mycobacteria had earlier been studied by Barclay and Ratledge [61] for their mycobactins which were of particular interest as some freshly isolated strains of *M. avium* had been reported as being dependent on mycobactin for growth [62]. The presence of mycobactins in strains of *M. avium* had, though, been initially reported by Ratledge and McCready [63]. Barclay and Ratledge [61] found that only those strains of *M. avium* that could grow with-out mycobactin could produce it themselves but three strains, initially unable to grow unless mycobactin was added to the growth medium, were eventually able to grow without it and now produced small quantities themselves. This indi-cated that mycobactin biosynthesis was being strongly repressed and then slowly reversed during the subsequent adaptation rather than being indicative of a perma-nent genetic deletion. The structure of mycobactin Av from *M. avium* and other MIAS strains was subsequently determined by Barclay et al. [64] (Fig. 2.4). In addition, the same type of mycobactin was also isolated from three strains of *M.*

Table 2.2 Occurrence of mycobactins in *Mycobacterium* and related species[a]

Species in which mycobactin has been found in all strains examined	***Mycobacterium***: M. aurum, M. avium[b], M. bovis BCG, M.chelonae, M. chitae, M. diernhoferi, M. duvali, M. farcinogenes, M. flavescens, M. fortuitum, M. gadium, M. gordonae, M. gallinarum, M. intracellulare, M. kansasii, M. marinum, M. neoaurum, M. nonchromogenicum, M. peregrinum, M. phlei, M. scrofulaceum, M. senegalense, M. smegmatis, M. szulgai, M. terrae, M. trivale, M. tuberculosis, plus unspeciated species termed "Armadillo-derived mycobacteria (ADM)".
	Nocardia: N. asteroides, N. brasilienses, N. caviae, N. farcinica
	Rhodococcus: R. bronchialis, R. rubropertinctus, R. terrae
Species in which mycobactin has been found in low (<0.1 %) concentration or is absent in some strains	M. parafortuitum, M. thermoresistible, M. vaccae
Species in which mycobactin has not been found	***Mycobacterium***. "M. kanazawa"[c], "M. komossense"[c], M. paratuberculosis[d]
	Rhodococcus: R. coprophilus, R. equi, R. erythropolis, R. Luteus, R. maris, R. rhodnii, R. rhodochrous, R. ruber

[a]Data from [26, 54–57, 59, 60, 71]
[b]A few strains are dependent upon mycobactin for growth
[c]May not be a true *Mycobacterium*
[d]One strain (adapted to laboratory growth) produces mycobactin J [64, 66]

paratuberculosis of that had lost their original dependency on mycobactin and could now produce it themselves. This suggested a taxonomic similarity between *M. paratuberculosis* and the MIAS complex of mycobacteria and this indeed has now been confirmed by more conventional taxonomic methods [65]. *M. paratuberculosis*, having started life as *Mycobacterium enteriditis chronicae pseudotuberculosae bovis*, Johne, or Johne's bacillus for short, and then becoming *M. johnei*, is now re-named as *M. avium* subsp. *paratuberculosis*.

Barclay et al. [64] found that mycobactin Av differed from the other mycobactins in having two long lipophilic side-chains (Fig. 2.4), one at the usual R_1 position and the second one of about 10 carbons in length at the R_4 position. They also showed that the mycobactins from *M. tuberculosis*, *M. bovis* and *M. africanum*, were identical molecules and therefore could all be named as mycobactin T.

A mycobactin from M. *paratuberculosis*, but of slightly different structure to that isolated by Barclay et al. [64], had been isolated and identified 2 years earlier by Richard Merkal [66, 67]. Again this was being produced by a strain that was initially mycobactin-dependent for growth but had subsequently reverted. This was named as mycobactin J (Fig. 2.4) but it was more like the conventional mycobactins in having just a single alkyl chain at the R1 position. The absolute configuration of the structure of mycobactin J was subsequently confirmed by Schwartz and De Voss [68]. Barclay et al. [64], having kindly been sent a sample of mycobactin J and also the production organism, *M. paratuberculosis* strain NADC 18 (now

ATCC 19698) by Richard Merkal, could only find mycobactin J as a minor component at <10 % of the total mycobactins that they extracted from the same strain as used by Merkal himself. The major mycobactin of this species, in our hands, was equivalent to the mycobactin from *M. intracellulare* M12, which is part of the avium complex of mycobacteria. It now appears that the original culture of *M. paratuberculosis* strain 18 probably had been taxonomically misnamed [69, 70] and should therefore be regarded as a strain of *M. avium*.

What the reasons were for mycobactin J being the dominant siderophore when NADC 18 strain was grown in Merkal's laboratory but not in the UK remains unsolved but clearly there must be some subtle regulatory mechanism that can cause such a shift. Mycobactin J remains the only commercially available mycobactin and, because of its structural similarities to the main mycobactins, is able to stimulate the growth of both a number of isolates and mutants that are being generated with defects in their iron metabolic pathways.

Hall and Ratledge [71] also isolated mycobactin-like materials from three out of 11 species of *Rhodococcus*: *R. bronchialis*, *R. terreus* and *R. rubropertinctus* and suggested that the latter two species might be equivalent to each other. Although no structural determinations were carried out, the similarity of these new materials to the mycobactins was apparent and they should have been, but were not, named as rhodobactins. Some species of *Rhodococcus*, including *R. rhodochrous*, did not, however, produce a 'rhodobactin'. Some 20 years later, Dhungana et al. [72] isolated a siderophore from *R. rhodochrous*, albeit from a different strain to that of Hall and Ratledge, with hexadecane as the sole carbon source. This siderophore was named rhodobactin but it was present in the culture medium and was not apparently in the cells. It was therefore distinct in structure to the mycobactins having two catecholate and one hydroxamate moieties for iron chelation instead of having a salicyloyl and two hydroxamate moieties and without a long alkyl chain for lipid solubility. Pedantically, the name 'rhodobactin' is therefore incorrect as it would imply similarity to the mycobactins and nocobactins.

2.4.3 Salicylic Acid

The story of the mycobactins and iron metabolism was then taken up by the author of this review. I carried out research work, as post-doctoral fellow, with Frank Winder in Dublin from October 1960 to June 1964 being asked to focus on the metabolic consequences of iron-deficiency in *M. smegmatis* being used as a model organism for the tubercle bacillus. The first significant finding [73] was the identification of salicylic acid that accumulated up to about 17 μg/ml in the medium of iron deficiently grown cells whereas in iron replete medium only about 0.6 μg salicylic acid/ml was found (see Fig. 2.6). However, a mistake was made in the latter part of this paper when the concentrations of salicylic acid were being given for *M. tuberculosis* and *M. phlei*. The 'salicylic acid' in the latter species had only been verified by simple paper chromatography where it ran with the same Rf value as authentic

Fig. 2.6 The accumulation
of salicylic acid during
iron-deficient growth of
Mycobacterium smegmatis.
Light bars iron-sufficient
growth conditions (controls)
and *dark bars* iron-deficient
growth [73, 113]

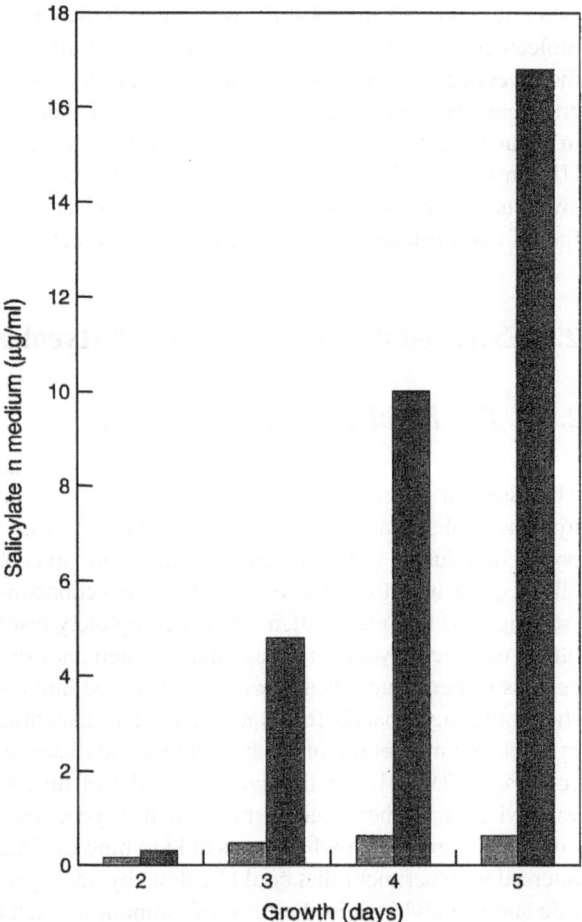

salicylic acid. Alan Snow then wrote to us to ask if the detected acid was indeed salicylate; could it not be 6-methylsalicylate? And, indeed, so it turned out to be.

Snow's concern over the correct identification of salicylic acid arose from his realization that, although salicylate was a common moiety in most mycobactins, in some mycobactins (see Fig. 2.4) and notably in mycobactin P from *M phlei*, it was 6-methylsalicylate that occurred. It therefore seemed odd to Snow that we had recorded finding salicylate in the extracellular medium of this species as this would imply a metabolic puzzle. However, by using more selective solvent systems for paper chromatography, we then showed that the original conclusion had been too hasty. We had failed to double-check; *M. phlei* did indeed secrete 6-methylsalicylate into the culture medium and not salicylic acid.

This useful contact with Alan Snow then immediately brought our attention to the mycobactins which, at that point, were just emerging as probably having significant roles in iron metabolism. Although the biochemical connection between salicylic acid and mycobactin was clear, it was far from evident how these two molecules might interact, if at all, to achieve uptake and assimilation of iron into the mycobacteria. However, for a continuation of this work, there was jump of several years as, following my time in Dublin, I then spent the next 3 years working in industry and it was only when I returned to academia to the University of Hull, UK, in late 1967 that interest in the mycobacteria and iron metabolism was re-awakened. The role of salicylic acid in mycobacteria then became one of the major foci of our work and this is described in more detail in the following section.

2.5 Extracellular Siderophores of Mycobacteria

2.5.1 The Problem with Mycobactin

The essential puzzle with mycobactin was not whether or not it was involved with iron metabolism, as clearly it was just judging from the very large increase in its production during iron-deficient growth of any mycobacterium, but how did it actually acquire iron from the environment. Mycobactins all have a long alkyl chain (see Fig. 2.4) that makes them almost completely insoluble in water but easily soluble in organic solvents such as ethanol, methanol or chloroform. Thus, they were cell-associated materials and were not released into the culture medium. It was later shown that mycobactin forms a discrete but discontinuous layer abutting on to the cytoplasmic membrane of the cells and some distance from the outer surface of the cell (Fig. 2.7) [74]. For this work, staining of mycobactin was achieved by using vanadyl ions as they reacted faster with mycobactin than did iron and was also much more specific in what it was able to bind to. (Alan Snow pointed out to me an interesting experiment that could be done by adding a mixed solution of ferric chloride and vanadyl ions in the form of ammonium metavanadate to desferrimycobactin in ethanol. The solution would immediately turn deep blue due to formation of the vanadate complex but then, on standing overnight, the solution became red due to the formation of the ferric complex. This indicated a difference between the rate of the reaction and the stability of the product. Ferric ions although reacting slower than vanadyl ions nevertheless could replace it and form a complex with a much higher affinity which was then the final stable form of the chelate.)

It was suggested in this paper of Ratledge et al. [74] that the mycobactin was probably intercalated between the cytoplasmic membrane and the peptidoglycan backbone of the cell wall with, perhaps a small amount of it possibly being within the membrane itself. There did not appear to be any of it within the cytoplasm. Thus, what was the function of mycobactin and how was it able to acquire iron from the extracellular medium, or the environment in which it grew within an infected host cell, if it was a wholly intracellular and water-insoluble material? Ivan Kochan [75] working in Miami of Ohio University, suggested that mycobactin could indeed fulfill this role by demonstrating that mycobactin could remove

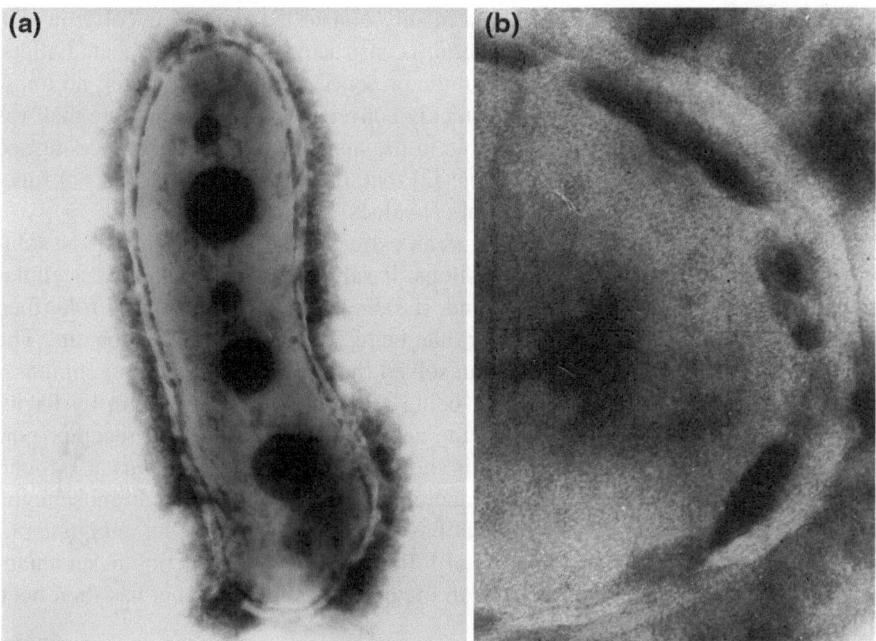

Fig. 2.7 **a** (magnification ~140,000-fold), and **b** (magnification ~700,000-fold). Cellular location of mycobactin. Electron micrographs of iron-deficiently grown Mycobacterium smegmatis incubated with 0.1 % ammonium metavanadate for 10 min at 4 °C (The vanadyl ions react very quickly and specifically with desferrimyobactin but are not metabolically removed from the mycobactin). The large *black circular* areas in the cytoplasm are polyphosphate granules which are also seen in iron-sufficiently grown cells [74]

the iron from transferrin but this did not explain how the two molecules might legitimately come into contact. Kochan et al. [76] and later Golden et al. [77] then showed that mycobactin could be solubilized and transferred into the medium if a detergent, such as Tween 80, Triton or lecithin, was added to the medium. But again, to my group (at Hull University), that now included Leo Macham as postdoctoral research assistant and who made some significant contributions in this area, this did not seem a likely condition for pathogenic mycobacteria to experience when growing in vivo. To us, the experimental conditions used by Kochan seemed contrived and unlikely to be realistic. His proposals certainly could not explain how iron was mobilized by mycobacteria growing in simple laboratory culture medium devoid of any detergent. We, therefore, were of the opinion that another iron binding component, that was water-soluble and which would be released by the cells into their surrounding environment, was necessary to explain how iron was solubilized in the medium before being transferred into the cells.

The first and obvious candidate for this was salicylic acid—why else did it occur in increased quantities in the culture medium during iron-deficient growth (Fig. 2.6)?

Although [55]Fe-labelled salicylate could readily donate the iron to mycobactin [78] this was only if the experiment was carefully constructed to avoid phosphate buffers. When similar experiments were done in the presence of phosphate buffer, no transfer of iron took place as the iron was quickly converted to ferric phosphate that was insoluble and not immediately accessible to the mycobactin [79]. It was confirmed much later by Chipperfield and Ratledge [2] that, indeed, salicylate could not function as chelating agent for iron at neutral pH values.

The failure of salicylate to function as an extracellular siderophore for the solubilization of iron then raised two questions: if salicylate was not the extracellular iron sequestering agent, what was? And, if salicylate did not have this role, then what function did it have, if any, other than being a precursor of mycobactin? The answer to the latter question remains unsolved though salicylate can function as a means of transferring the ferrous ion, being released from mycobactin by ferric-mycobactin reductase [38–40] across the cell membrane and into a receptor porphyrin for the synthesis of haem. This is shown later in Fig. 2.11. It was also found that the mode of action of the anti-tuberculosis agent, PAS = p-aminosalicylic acid, was, contrary to the initial indications that it was acting as an antagonist of p-aminobenzoic acid for the synthesis of folic acid, was in fact acting as an antagonist of salicylic acid for its role in iron metabolism [80, 81]. This has then been confirmed by later work [82–84].

An alternative role for salicylic acid was advanced by Morrison [42] who pointed out that salicylate may be acting as an energy-uncoupling agent as had been indicated earlier by Brodie [85]. This then might tie in with the original observation by Bernheim [86] that salicylate was readily oxidized by *M. tuberculosis* and therefore might be an important respiratory substrate. The view of Brodie [84, 87] would be that salicylate uncoupled oxidative phosphorylation and, therefore, with insufficient ATP for growth, cells have to increase their respiration rate. But in the iron-deficient cells that over-produce salicylate, the salicylate may be a means of rapidly down-regulating energy production by directly uncoupling oxidative phosphorylation in order to conserve key metabolic processes and not to generate ATP needlessly. This is certainly an intriguing suggestion from Morrison that has never been followed up.

The search for the iron-solubilizing agent that would be present in the culture medium of mycobacteria then began. Clearly the answer to this question was of considerable importance as it would help to explain how pathogenic mycobacteria were able to acquire iron from host tissues and iron-containing cell components. Macham and Ratledge [88] carried out experiments with *M. smegmatis* and *M. bovis* BCG both being grown iron deficiently. They showed that there was some material in the cell-free culture filtrates of both organisms that could hold [55]Fe in solution at pH 7 in the presence of phosphate ions and could be dialyzed; in other words, it had a molecular size of <10000 Da and was therefore not some form of colloidal iron. These materials were named exochelins. But it quickly became evident that the material from *M. smegmatis* was quite different from that of *M. bovis*. The former was completely water-soluble and could not be extracted into any organic solvent, including ethanol, whereas the other exochelin could, when in the ferric-form, be extracted into chloroform [88, 89]. It was also found that the two exochelins had different mechanisms for iron uptake: exochelin from *M. smegmatis* was taken up by an active (energy-dependent) transport system but the one from *M. bovis* was by facilitated diffusion and was not energy dependent [90, 91]. Furthermore, the *M. smegmatis* exochelin could not be taken up by *M. bovis* but that from *M. bovis* could be taken up by *M. smegmatis*. Thus, two distinct types of extracellular siderophores were being produced by the mycobacteria. Other water-soluble exochelins were recovered from other non-pathogens: *M. neoaurum* [92] and *M. vaccae* [93]. The latter siderophore was of interest as this species of mycobacteria did not appear to have a mycobactin.

2.5.2 The Water-Soluble Exochelins

It was nearly 20 years after the initial discoveries of the exochelins before their structures were resolved though it was established soon after their initial isolation that exochelin MS from *M. smegmatis* was probably a pentapeptide with three *N-epsilon*-hydroxyornithines providing the chelating centre. Various research groups in the UK had been approached by the author for assistance in trying to work out the structures and to see how the ornithine residues, together with a *beta*-alanine and an *allo*-threonine, were assembled. But none had been able to complete the work until the author approached the group headed by Dudley Williams in the Department of Chemistry at the University of Cambridge to help solve the structure of the water-soluble exochelins. The structure of the chloroform-soluble exochelins was solved by collaboration with a research group at Glaxo Research Laboratories.

The structure of the exochelin from *M. smegmatis*, then named as exochelin MS, was determined by a PhD student, Gary Sharman working under Dudley Williams at Cambridge University. Exochelin MS was an ornithinyl siderophore with three hydroxamate groups that provided the iron chelating center (Fig. 2.8a) [94]. Further, but unpublished, work of the author indicated that the exochelin from *M. vaccae* was probably similar if not identical to this exochelin. The structure of the exochelin from *M. neoarum*, called exochelin MN, was different [95] but was still based on a peptide backbone as seen with exochelin MS (Fig. 2.8b). It had an unusual 2-hydroxyhistidine residue as part of its iron chelating center.

Although the structure of exochelin MS was not elucidated until 1995, quite a lot of its properties and function had been worked out beforehand. Its uptake, see above, was by an active transport process requiring the input of energy (i.e., ATP was involved at some point of the mechanism) [90]. It was produced in a growth-related manner and could readily solubilize iron from not only inorganic forms of insoluble iron, such as ferric hydroxide and ferric phosphate, but also from ferritin the storage form of iron found in all animals [88, 89]. The involvement of mycobactin in the uptake of iron into *M. smegmatis* was not immediately apparent when small concentrations of ferric-exochelin were used; however, when higher concentrations were used, a second uptake process became evident. This was a slower

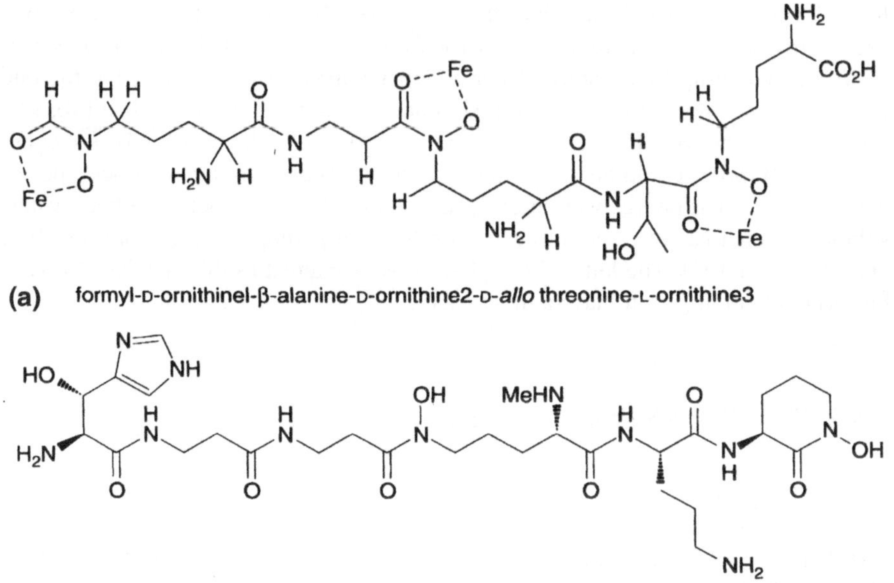

(a)   formyl-D-ornithine1-β-alanine-D-ornithine2-D-*allo* threonine-L-ornithine3

(b) L-*threo*-β-hydroxy histidine-β-alanine-β-alanine-L-α methyl ornithine-L-ornithine-L-(cyclo)ornithine

Fig. 2.8 Structures of the exochelins (water-soluble extracellular siderophores) from **a**
Mycobacterium smegmatis, and **b** *M. neoaurum* [94, 95]

process and was not inhibited by energy poisons and other agents; it was deduced to
be by a process involving transfer to mycobactin itself [90] and could be explained
by the mycobactin preventing a sudden over-load of iron into the cells. Cells that
were iron-deficient had very low contents of porphyrins (Table 2.1) and thus the key
precursors of haem synthesis were not instantly available to utilize the iron. But even
though iron could not be immediately used and incorporated into cell components,
a mechanism of iron storage was necessary that would then serve as a 'pantry' of
iron. This then began to shape the view that mycobactin was an intracellular store of
iron and that it acquired iron only when there was a sudden availability of it to the
cells (see Fig. 2.11). Of interest was the finding that an exochelin may be involved
in iron uptake into the leprosy bacillus, *M. leprae.* Somewhat fortuitously, Hall et
al. [96] had isolated the exochelin from *M. neoarum* simply because this species
was taxonomically related to *M. vaccae* [97] and which, in turn, had been suggested
might be related to *M. leprae* [98]. Iron metabolism in *M. neoarum* therefore might
be worth investigating. Richard Hall showed that, by using [55]Fe-labelled exochelin
MN, the iron was taken up by cells of *M. leprae* isolated from armadillo livers but
the process was not one of active transport (as was with its uptake into *M. neoarum*)
and appeared to be by facilitated diffusion [96]. The process though was specific in
that there was no transfer of iron when chelated to exochelin MS. However, another
exochelin, this time isolated from an armadillo-derived *Mycobacterium* (ADM)
and which could be grown in the laboratory, also could donate iron to *M. leprae*

[92]. Examination of the two exochelins—from *M. neoaurum* and *Mycobacterium* ADM8563—showed that they had similar properties and were possibly identical. The presence of a chloroform-soluble carboxymycobactin was not investigated in this work but it would not have been present in the preparations used because these had been purified by ion exchange chromatography which would have excluded the carboxymycobactins. The rate of uptake of iron into *M. leprae* was, as might have been expected, very slow but, as exochelin-mediated iron uptake did not occur in ADM cells that had been grown iron sufficiently, it was concluded that this result might indicate that *M. leprae* had been growing iron-deficiently in its host animal (the armadillo) in order for iron uptake to have taken place at all. It was also suggested that one of these exochelins might be useful additions to any growth medium that might be being developed for the possible growth of *M. leprae* in the laboratory. This however was never additionally studied and the cultivation of *M. leprae* in culture medium still remains a distant prospect.

2.5.3 The Carboxymycobactins

The name 'carboxymycobactin' was not given to the extracellular siderophores that had been isolated from pathogenic species of mycobacteria, including *M. tuberculosis, M. bovis* and *M. avium* and related species, before their structures had been established which was not until 1995. Up to that date, they were referred to as the chloroform-soluble exochelins. Perhaps, with hindsight an alternative name to 'exochelin' might have been used to avoid confusion with the obviously different water-soluble exochelins that were described in the previous section. But we (as all the work had been done in the author's laboratory) simply had no idea what might be their structures. To call the material 'exomycobactin', as was belatedly suggested by another group, presupposed we knew it was related to mycobactin itself. For all we knew, it could have been based on a completely different type of structure. What made matters worse, was that initial analysis of the siderophores from the pathogenic species showed that they were composed of a variety of materials. Barclay and Ratledge [48, 99] analyzed the exochelins from species of mycobacteria belonging to the tuberculosis group, including both H37Rv (virulent) and H37Ra (avirulent) strains of *M. tuberculosis*, and to the *avium* groups using both high performance thin layer chromatography (HPTLC) and high performance liquid chromatography (HPLC). Multiple spots or peaks were revealed; in some cases upwards of 15 individual compounds could be seen. This was completely puzzling and suggested that, as there was no large single entity, determining the structures might be a long and daunting task if each component had to be isolated and purified. If only we had used a less discriminating technique, such as ordinary thin layer chromatography, we might have seen just a single spot on the chromatograms that would then have encouraged us to have the material examined without delay. But a multiple of spots on HPTLC and peaks with HPLC suggested there might be a multiple of structures.

It was, however, quite clear to us from very early in our experiments with the chloroform-soluble exochelins that they were of significance in the uptake of iron into pathogenic mycobacteria. Perhaps the most important work on their role came from the work of Raymond Barclay who was a post-doctoral research assistant working with me. Raymond showed that the bacteriostatic effects of serum (containing the iron-withholding protein of transferrin) towards the growth of *M. avium* and *M. paratuberculosis* in laboratory medium could be reversed not only by mycobactin but also by the exochelins that these bacteria were producing [48]. This was strong evidence in favor of these siderophores being of major importance in the development of mycobacterial infections in animals. Our much earlier work following the initial discovery of the exochelins had already established their ability to extract the iron from animal ferritin to support growth of *M. bovis* var. BCG [100]. Finding out what was the structure or structures of these siderophores then became major of pre-occupation in our work. But would be nearly another 10 years before the problem was finally solved.

The breakthrough in determining the structure of the chloroform-soluble exochelins came in collaboration between researchers at Glaxo Research Laboratories at Stevenage near London. As soon as the first high resolution NMR spectrum of the purified exochelin from *M. avium* was obtained, Steve Lane and his colleagues were immediately able to suggest a strong similarity in structure to mycobactin but with a variation: the long alkyl chain of mycobactin was now shorter and, instead of terminating in a methyl group, now terminated in a carboxy group (Fig. 2.9). As there was now competition to publish the structure of these extracellular siderophores, an instant decision was made to call the material 'carboxymycobactin' as this seemed to be an accurate descriptor of the molecule. The structure was then published by Lane et al. [101] with the paper being submitted on February 15th 1995. In this paper, mention was made that a preliminary examination of the related materials from *M. tuberculosis* and *M. bovis* had revealed that these were also carboxymycobactins of similar structures. A group led by Marcus Horwitz in California, USA, simultaneously published (their paper was submitted on February 21st, a week after the paper of Lane et al. [101]) the structure

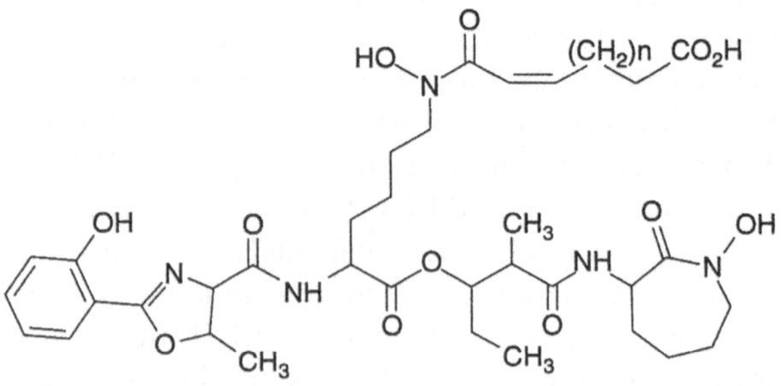

Fig. 2.9 Structure of the carboxymycobactins (originally termed the chloroform-soluble exochelins) from *Mycobacterium avium, M. bovis* BCG and *M. tuberculosis* [101] where n = 2–9. Related siderophores have been reported in *M. smegmatis* [53, 104]. Related molecules but with a terminal methyl ester group on the acyl chain from *M. tuberculosis* and *M. avium* were found [102, 103]

of the related material that they had isolated from *M. tuberculosis* [102, 103]. In this case, however, the terminating group of the alkyl chain was not a carboxy group but was the methyl ester of this group. Possibly the esterification reaction may have occurred during the late growth phase of the cultures that had been used whereas for the work of Lane et al. [101, 104], culture filtrates had been prepared from cells in their active phase of growth during which iron solubilization and uptake should be at their maximum.

It is clear that all the siderophores obtained earlier from culture filtrates of pathogenic mycobacteria and had been termed 'chloroform-soluble exochelins' were, in fact, carboxymycobactins. A list of species and strains that produce carboxymycobactins is given in Table 2.3. Of considerable importance and interest was the finding that all (13 out of 13) strains of *M. paratuberculosis* that had been examined by Barclay and Ratledge [99] produced a carboxymycobactin but all of them still required mycobactin for growth. However, it was ruled out that the added mycobactin was being converted into the extracellular product as the amount of carboxymycobactin recovered was much greater than the amount of mycobactin that had been added as a growth factor into the culture medium. Also there was no observable conversion of ^{14}C-labelled carboxymycobactin into mycobactin itself when fed to cultures of *M. bovis* [100].

This discovery then provides a partial explanation as to how *M. paratuberculosis* and other mycobactin-dependent species are able to grow in vivo. They simply produce carboxymycobactin, an extracellular siderophore that then acquires the iron from host tissues and transfers the iron to the bacterial cells. There must then be a slow transfer of iron into the pathogenic bacteria without the participation of mycobactin, though how this occurs is still not clear. Evidently, however, when the bacteria are isolated from an infected animal, mycobactin becomes essential for the uptake process to be complete and for the bacteria to grow in vitro. Homuth et al. [105] took an alternative view: they suggested, on the basis of their finding of an extracellular ferric reductase in cultures of *M. paratuberculosis* that could reduce the iron in transferrin and lactoferrin, that this may be an alternative means of iron acquisition in this bacterium in vivo. The possibility of the direct uptake of iron from transferrin had been suggested earlier by Lambrecht and Collins [106]. But, if this were the case, one would then have expected extracts of animal tissues to have supported growth of this bacterium in laboratory culture medium but clearly this does not happen [22–24]. Further work on this aspect of iron metabolism in *M. paratuberculosis* and other mycobactin-dependent strains is therefore still needed to give some vital clues as to how it is accomplished.

Table 2.3 Occurrence of carboxymycobactins in mycobacteria (see also Fig. 2.9)[a]

Pathogens	*M. africanum, M. avium, M. bovis* BCG, *M. intracellulare, M. paratuberculosis, M. trivial, M. tuberculosis* H37Ra and H37Rv, *M. xenopi*
Non-pathogens	*M. smegmatis*[b] *M. neoaurum*[c]

[a]Data from [60, 99]
[b][108]
[c]From Tan Eng Lee and Ratledge [unpublished]

The only species of *Mycobacterium* that failed to produce detectable amounts of carboxymycobactin was *M. microti* which also appeared not to produce a myco-bactin [99]. This species therefore is an oddity amongst the mycobacteria in pro-ducing neither mycobactin nor a carboxymycobactin and clearly would warrant further investigation as to what is its mechanism of iron acquisition. It is, however, a genuine mycobacterium in that it is the causative organism of tuberculosis in voles and other rodents but has also been associated with some isolated cases of TB in humans [107]. Some investigation into iron metabolism in this species may therefore be warranted.

What came as a complete surprise during this phase of our work was the find-ing of a carboxymycobactin in culture filtrates of *M. smegmatis* grown iron defi-ciently [108]. The presence of exochelin MS in this species had appeared to be sufficient to account for iron uptake in the saprophyte so the presence of a second, and alternative, method of iron acquisition was unexpected. However, the amounts of carboxymycobactin (10–25 µg/ml) that were produced were, at most, only 10 % of the total siderophores. Also it was produced later in the growth of the cells indicating that the exochelin-dependent route of iron assimilation was probably the major one. It was also noted that the amount of carboxymycobactin was some 20 times higher when glycerol was used as a carbon source instead of glucose. The structure of this carboxymycobactin was subsequently determined [104] and was found to resemble the structure of mycobactin S with a family of short car-boxylic acids attached to the mycobactin nucleus. Some eight variants of the mol-ecule were identified [6, 104] and this could then partly explain why Barclay and Ratledge [61, 99] found such a plethora of iron-binding molecules when they first examined the carboxymycobactins from pathogenic mycobacteria by high resolu-tion chromatographic techniques.

2.6 Putting it All Together

How the various components of iron metabolism come together to present a coher-ent picture of iron uptake and transport in mycobacteria still has a long way to go. But we are getting there. It is not, however, the remit of this chapter to go into the details of how the various components enumerated above dovetail together but, for the sake of completeness, I have included the very first model that was proposed in 1975 for iron transport in the mycobacteria (Fig. 2.10) that does not distinguish between the water-soluble exochelins of the saprophytic strains and the chloroform-soluble ones of the pathogens [100]. A more detailed model proposed in 1999 is given in Fig. 2.11. In this model, besides the major routes of iron uptake via the exochelins and carboxymycobactins, the uptake of iron chelated to citrate was worked out by Ann Messenger [93] and assimilation of iron by direct binding of a mycobacterial cell to transferrin or lactoferrin was proposed by Lambrecht and Collins [106] probably involving an extracellular ferric reductase as iden-tified by Homuth et al. [105]. This enzyme could also work with ferritin and

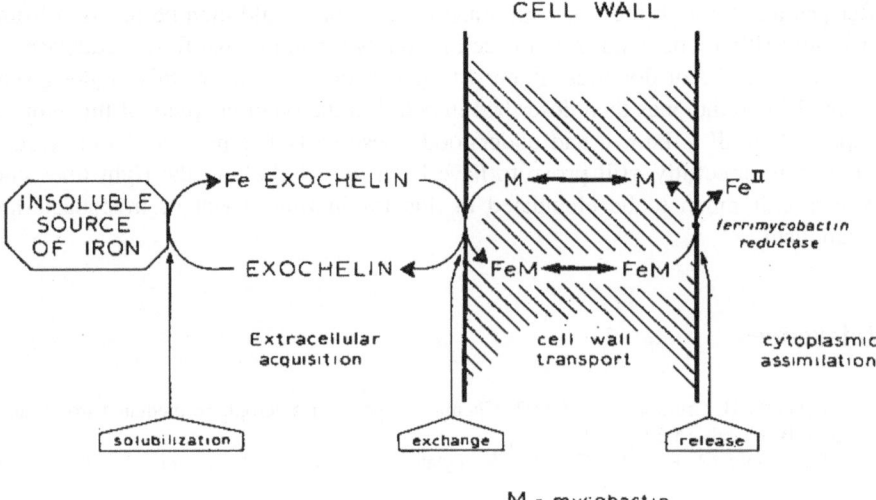

Fig. 2.10 An early model proposed to explain iron uptake and transport in the mycobacteria [98]

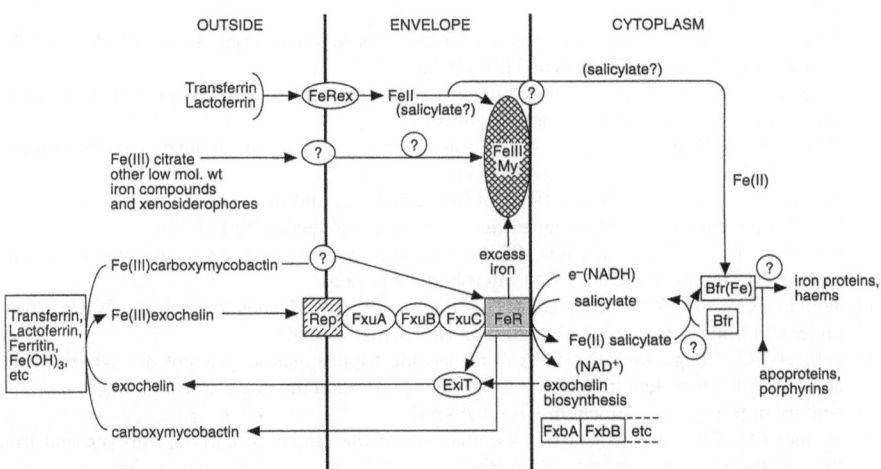

Fig. 2.11 Possible mechanisms for iron uptake in mycobacteria as further elucidated during the 1980s and 1990s [113]. *FeRex* an extracellular ferric reductase, *FeR* ferric-mycobactin reductase, *My* mycobactin, *Rep* receptor protein for exochelin, *FxuA, FxuB* etc. ferric-exochelin uptake proteins, *ExiT* exochelin transport protein, *FxbA, FxbB*, etc. ferric-exochelin biosynthesis proteins, *Bfr* bacterioferritin, *?* unknown mechanisms but the uptake of ferric-carboxymycobactin is now known to involve an ABC transporter

ferric ammonium citrate. Iron storage within the cytoplasm of the cell is by bacterioferritin [109–111] but only when the iron is in excess of immediate metabolic requirements. Bacterioferritin receives the iron being released by ferrimycobactin after passage through the cell membrane [112]. Iron would then be released from bacterioferritin on demand from the cell probably by an internal ferric-reductase.

It will then be of doubtless interest to the reader to compare this model given in Fig. 2.11 to the ones then described in detail in the other chapters of this monograph. When all is known and understood, these early beginnings of iron acquisition may, hopefully, still prove to have been modeled along the right lines but there is still much to be elucidated in this fascinating aspect of mycobacterial metabolism.

References

1. Boukhalfa H, Crumbliss AL (2002) Chemical aspects of siderophore mediated iron transport. Biometals 15:325–339
2. Chipperfield JR, Ratledge C (2000) Salicylate is not a bacterial siderophore: a theoretical study. Biometals 13:165–168
3. Sauton A (1912) Sur la nutrition minerale du bacilli tuberculeux. CR Acad Sci (Paris) 155:860–861
4. Edson NL, Hunter GJE (1943) Respiration and nutritional requirements of certain members of the genus *Mycobacterium*. Biochem J 37:563–571
5. Goth A (1945) The antitubercular activity of aspergillic acid and its probable mode of action. J Lab Clin Med 30:899–905
6. Turian G (1951) Action plasmogene du fer chez les Mycobacteries. Le bacilli de la fleole, indicateur du fer. Helv Chim Acta 34:917–920
7. Winder F, Denneny J (1959) Effect of iron and zinc on nucleic acid and protein synthesis in *Mycobacterium smegmatis*. Nature 184:742–743
8. Winder FG, O'Hara C (1962) Effect of iron deficiency and of zinc deficiency on the composition of *Mycobacterium smegmatis*. Biochem J 82:98–102
9. Winder FG, O'Hara C (1964) Effects of iron deficiency and of zinc deficiency on the activities of some enzymes in *Mycobacterium smegmatis*. Biochem J 90:122–126
10. Donald C, Passey BI, Swaby RJ (1952) A comparison of methods for removing trace metals from microbiological media. J Gen Microbiol 7:211–220
11. Winder FG, O'Hara C (1966) Levels of iron and zinc in *Mycobacterium smegmatis* grown under conditions of trace metal limitation. Biochem J 100:38P
12. Winder FG, Coughlan MP (1969) A nucleoside triphosphate-dependent deoxyribonucleic acid-breakdown system in *Mycobacterium smegmatis* and the effect of iron limitation on the activity of this system. Biochem J 111:679–687
13. Winder FG, Coughlan MP (1971) Comparison of the effects of carbon, nitrogen and iron limitation on the growth and on the RNA and DNA content of *Mycobacterium smegmatis*. Irish J Med Sci 140:16–25
14. Winder FG, McNulty MS (1970) Increased DNA polymerase activity accompanying decreased DNA content in iron-deficient *Mycobacterium smegmatis*. Biochim Biophys Acta 209:578–580
15. Winder FG, Lavin MF (1971) Partial purification and properties of a nucleoside triphosphate-dependent deoxyribonuclease from *Mycobacterium smegmatis*. Biochim Biophys Acta 247:542–561
16. Winder FG, Sastry PA (1971) The formation of a long-lived complex between an ATP-dependent deoxyribonuclease and DNA. FEBS Lett 17:27–30
17. Winder FG, Barber DS (1973) Effects of hydroxyurea, nalidixic acid and zinc limitation on DNA polymerase and ATP-dependent deoxyribonuclease activities of *Mycobacterium smegmatis*. J Gen Microbiol 76:189–196

18. Ehrenberg A, Reichard P (1972) Electron spin resonance of the iron-containing protein B2 from ribonucleotide reductase. J Biol Chem 247:3485–3488
19. MacNaughton AW, Winder FG (1977) Increased DNA polymerase and ATP-dependent deoxyribonuclease activities following DNA damage in *Mycobacterium smegmatis*. Mol Gen Genet 150:301–308
20. McCready KA, Ratledge C (1978) Amounts of iron, haem and related compounds in *Mycobacterium smegmatis* grown in various concentrations of iron. Biochem Soc Trans 6:421–423
21. McCready KA (1980) Studies on iron metabolism in *Mycobacterium smegmatis* and other mycobacteria. PhD thesis, University of Hull, UK
22. Twort FW, Ingram GLY (1912) A method for isolating and cultivating the *Mycobacterium enteritidis chronicae pseudotuberculosae bovis*, Johne, and some experiments on the preparation of a diagnostic vaccine for pseudo-tuberculosis enteritis of bovines. Proc R Soc Lond B Biol Sci 84:517–530
23. Twort FW, Ingram GLY (1913) A monograph on Johne's disease (enteritis chronicae pseudotuberculosa bovis). Baliere, Tindall and Cox
24. Twort and Ingram (1914) Further experiments on the biology of Johne's bacillus. Zentr Bakteriol Parasitenk Abt 1 Org 73:277–283
25. Johne HA, Frothingham L (1895) Ein eigentuemlicher Fall von Tuberculose beim Rind. Dtsch Ztschr Tiermed Pathol 21:438–454
26. Snow GA (1970) Mycobactins: iron-chelating growth factors from mycobacteria. Bacteriol Rev 34:99–125
27. Francis J, Madinaveitia J, Macturk HM Snow GA (1949) Isolation from acid-fast bacteria of a growth-factor for *Mycobacterium johnei* and of a precursor of phthiocol. Nature 163:365–366
28. Francis J, Macturk HM, Madinaveitia J Snow GA (1953) Mycobactin, a growth factor for *Mycobacterium johnei*. I isolation from Mycobacterium phlei. Biochem J 55:596–607
29. Wheater DWF, Snow GA (1966) Assay of the mycobactins by measurement of the growth of Mycobacterium johnei. Biochem J 100:47–49
30. Antoine AD, Morrison NE, Hanks JH (1964) Specificity of improved methods for mycobactin bioassay by *Arthrobacter terregens*. J Bacteriol 88:1672–1677
31. Reich CV, Hanks JH (1964) Use of *Arthrobacter terregens* for bioassay of mycobactin. J Bacteriol 87:1317–1320
32. Snow GA (1954a) Mycobactin. A growth factor for *Mycobacterium johnei*. II Degradation and identification of fragments. J Chem Soc 2588–2596
33. Snow GA (1954b) Mycobactin. A growth factor for *Mycobacterium johnei*. III Degradation and tentative structure. J Chem Soc 4080–4093
34. Snow GA (1961) An iron-containing growth factor from *Mycobacterium tuberculosis*. Biochem J 81:hn 4P
35. Snow GA (1965) The structure of mycobactin P, a growth factor for *Mycobacterium johnei*, and the significance of its iron complex. Biochem J 94:160–165
36. Snow GA (1965) Isolation and structure of mycobactin T, a growth factor from *Mycobacterium tuberculosis*. Biochem J 94:166–175
37. Hough E, Rogers D (1784) The crystal structure of ferrimycobactin P, a growth factor for the mycobacteria. Biochem Biophys Res Commun 57:73–77
38. Brown KA, Ratledge C (1974) Iron transport in *Mycobacterium smegmatis*: ferrimycobactin reductase [NAD(P)H:ferrimycobactin oxido-reductase], the enzyme releasing iron from its carrier. FEBS Lett 53:262–266
39. McCready KA, Ratledge C (1979) Ferrimycobactin reductase activity from *Mycobacterium smegmatis*. J Gen Microbiol 113:67–72
40. Ratledge C (1971) Transport of iron by mycobactin in *Mycobacterium smegmatis*. Biochem Biophys Res Commun 45:856–862
41. Morrison NE (1965) Circumvention of the mycobactin requirement of *Mycobacterium paratuberculosis*. J Bacteriol 89:762–767

42. Morrison NE (1995) *Mycobacterium leprae*: iron nutrition: bacterioferritin, mycobactin, exochelin and intracellular growth. Int J Lepr 63:86–91
43. White AJ, Snow GA (1969) Isolation of mycobactins from various mycobacteria. The properties of mycobactins S and H. Biochem J 111:785–792
44. Snow GA, White AJ (1969) Chemical and biological properties of mycobactins isolated from various mycobacteria. Biochem J 115:1031–1045
45. Snow GA (1969) Metal complexes of mycobactin P and of desferrisideramines. Biochem J 115:119–205
46. D'Arcy Hart P (1958) A mycobactin-containing liquid medium for the study of *Mycobacterium johnei*. J Pathol Bacteriol 76:205–210
47. Lambrecht RS, Collins MT (1992) *Mycobacterium paratuberculosis* factors that influence mycobactin dependence. Diagn Microbiol Infect Dis 15:239–246
48. Barclay R, Ratledge C (1986) Participation of iron on the growth inhibition of pathogenic strains of *Mycobacterium avium* and *M. paratuberculosis* in serum. Zentralbl Bakteriol Mikrobiol Hyg (Ser A) 262:189–194
49. Patel PV, Ratledge C (1973) Isolation of lipid-soluble compounds that bind ferric ions from *Nocardia* species. Biochem Soc Trans 1:886–888
50. Ratledge C, Snow GA (1974) Isolation and structure of nocobactin NA, a lipid-soluble iron-binding compound from *Nocardia asteroides*. Biochem J 139:407–413
51. Ratledge C, Patel PV (1976) Lipid-soluble, iron-binding compounds in *Nocardia* and related organisms. In: Goodfellow M, Brownell GH, Serrano JA (eds) The biology of the Nocardiae. Academic Press, London
52. Hall RM, Ratledge C (1982) A simple method for the production of mycobactin, the lipid-soluble siderophore from mycobacteria. FEMS Microbiol Lett 15:133–136
53. Ratledge C, Ewing DE (1978) The separation of the mycobactins from *Mycobacterium smegmatis* by using high-pressure liquid chromatography. Biochem J 175:853–857
54. Bosne S, Papa F, Clavel-Seres S, Rastogi N (1993) A simple and reliable EDDA method for mycobactin production in mycobacteria: optimal conditions and use in mycobacterial speciation. Curr Microbiol 26:353–358
55. Bosne S, Levy-Frebault VV (1992) Mycobactin analysis as an aid for the identification of *Mycobacterium fortuitum* and *Mycobacterium chelonae* subspecies. J Clin Microbiol 30:1225–1231
56. Hall RM, Ratledge C (1984) Mycobactins as chemotaxonomic characters for some rapidly growing mycobacteria. J Gen Microbiol 130:1883–1892
57. Leite CQF, Barreto AMW, Leite SRA (1995) Thin-layer chromatography of mycobactins and mycolic acids for the identification of clinical mycobacteria. Rev Microbiol 26:192–196
58. Barclay R, Furst V, Smith I (1992) A simple and rapid method for the detection and identification of mycobacteria using mycobactin. J Med Microbiol 37:286–290
59. Hall RM, Ratledge C (1985) Equivalance of mycobactins from *Mycobacterium senegalense*, *Mycobacterium farcinogenes* and *Mycobacterium fortuitum*. J Gen Microbiol 131:1691–1996
60. Hall RM, Ratledge C (1985) Mycobactins in the classification and identification of armadillo-derived mycobacteria. FEMS Microbiol Lett 28:243–247
61. Barclay RM, Ratledge C (1983) Iron-binding compounds of *Mycobacterium avium*, *M. intracellulare*, *M. scrofulaceum*, and mycobactin-dependent *M. paratuberculosis* and *M. avium*. J Bacteriol 153:1138–1146
62. Matthews PRJ, McDiarmid A, Collins P, Brown A (1977) The dependence of some strains of *Mycobacterium avium* on mycobactin for initial and subsequent growth. J Med Microbiol 11:53–57
63. Ratledge C, McCready KA (1977) Mycobactins from *Mycobacterium avium*. Int J Syst Bacteriol 27:288–289
64. Barclay R, Ewing DF, Ratledge C (1985) Isolation, identification, and structural analysis of the mycobactins of *Mycobacterium avium*, *Mycobacterium intracellulare*, *Mycobacterium scrofulaceum* and *Mycobacterium paratuberculosis*. J Bacteriol 164:896–903

65. Thorel MF, Krichevsky M, Levy-Frebault VV (1990) Numerical taxonomy of mycobactin-dependent mycobacteria, emended description of *Mycobacterium avium*, and description of *Mycobacterium avium* subsp. *paratuberculosis* subsp. nov., and *Mycobacterium avium* subsp. *silvaticum* subsp. nov. Int J Syst Bacteriol 40:254–260

66. Merkal RS, McCullough WG (1982) A new mycobactin, mycobactin J from *Mycobacterium paratuberculosis*. Curr Microbiol 7:333–335

67. McCullough WG, Merkal RS (1982) Structure of mycobactin. J Curr Microbiol 7:337–341

68. Schwartz BD, De Voss JJ (2001) Structure and absolute configuration of mycobactin. J Tetrahedron Lett 42:3653–3655

69. Collins DM, De Lisle GW (1986) Restriction endonuclease analysis of various strains of *Mycobacterium paratuberculosis* isolated from cattle. Am J Vet Res 10:2226–2228

70. Whipple DL, Le Febvre RB, Andrews RE et al (1987) Isolation and analysis of restriction endonuclease digestive patterns of chromosomal DNA from *Mycobacterium paratuberculosis* and other *Mycobacterium* species. J Clin Microbiol 25:1511–1515

71. Hall RM, Ratledge C (1986) Distribution and application of mycobactins for the characterization of species within the genus *Rhodococcus*. J Gen Microbiol 132:853–856

72. Dhungana S, Michalczyk R, Boukhalfa H et al (2007) Purification and characterization of rhodobactin: a mixed ligand siderophore from *Rhodococcus rhodochrous* strain OFS. Biometals 20:853–867

73. Ratledge C, Winder FG (1962) The accumulation of salicylic acid by mycobacteria during growth on an iron-deficient medium. Biochem J 84:501–506

74. Ratledge C, Patel PV, Mundy J (1982) Iron transport in *Mycobacterium smegmatis*: the locaction of mycobactin by electron microscopy. J Gen Microbiol 128:1559–1565

75. Kochan I, Pellis NR, Golden CA (1971) Mechanism of tuberculostasis in mammalian serum. Infect Immun 3:553–558

76. Kochan I, Cahall DL, Golden CA (1971) Employment of tuberculostasis in serum-agar medium for the study of production and activity of mycobactin. Infect Immun 4:130–137

77. Golden CA, Kochan I, Spriggs DR (1974) Role of mycobactin in the growth and virulence of tubercle bacilli. Infect Immun 9:34–40

78. Ratledge C, Marshall BJ (1972) Iron transport in *Mycobacterium smegmatis*: the role of mycobactin. Biochim Biophys Acta 279:58–74

79. Ratledge C, Macham LP, Brown KA et al (1974) Iron transport in *Mycobacterium smegmatis*: a restricted role for salicylic acid in the extracellular environment. Biochim Biophys Acta 372:39–51

80. Brown KA, Ratledge C (1972) Inhibition of mycobactin formation in *Mycobacterium smegmatis* by *p*-aminosalicylate: a new proposal for the mode of action of *p*-aminosalicylate. Am Rev Respir Dis 106:774–776

81. Brown KA, Ratledge C (1975) The effect of *p*-aminosalicylic acid on iron transport and assimilation in mycobacteria. Biochim Biophys Acta 385:207–220

82. Adilakshmi T, Ayling PD, Ratledge C (2000) Mutational analysis of a role for salicylic acid in iron metabolism of *Mycobacterium smegmatis*. J Bacteriol 182:264–271

83. Nagachar N, Ratledge C (2010) Roles of *trpE2*, *entC* and *entD* in salicylic acid biosynthesis in *Mycobacterium smegmatis*. FEMS Microbiol Lett 308:159–165

84. Nagachar N, Ratledge C (2010) Knocking out salicylate biosynthesis genes in *Mycobacterium smegmatis* induces hypersensitivity to *p*-aminosalicylate (PAS). FEMS Microbiol Lett 311:193–199

85. Brodie AF (1967) Microbial phosphorylating preparations: *Mycobacterium*. Methods Enzymol 10:157–169

86. Bernheim F (1951) Metabolism of aromatic compounds by mycobacteria. Adv Tuberc Res 5:5–39

87. Brodie AF, Gray CT (1957) Bacterial particles in oxidative phosphorylation. Science 125:534–537

88. Macham LP, Ratledge C (1975) A new group of water-soluble iron-binding compounds from mycobacteria: the exochelins. J Gen Microbiol 89:379–382

89. Macham LP, Stephenson MC, Ratledge C (1977) Iron transport in *Mycobacterium smegmatis*: the isolation, purification and function of exochelin MS. J Gen Microbiol 101:41–49

90. Stephenson MC, Ratledge C (1979) Iron transport in *Mycobacterium smegmatis*: uptake of iron from ferriexochelin. J Gen Microbiol 110:193–202

91. Stephenson MC, Ratledge C (1980) Specificity of exochelins for iron transport in three species of mycobacteria. J Gen Microbiol 116:521–523

92. Hall RM, Ratledge C (1987) Exochelin-mediated iron acquisition by the leprosy bacillus, *Mycobacterium leprae*. J Gen Microbiol 133:193–199

93. Messenger AJM, Hall RM, Ratledge C (1986) Iron uptake processes in *Mycobacterium vaccae* R877R, a mycobacterium lacking mycobactin. J Gen Microbiol 132:845–852

94. Sharman GJ, Williams DH, Ewing DF et al (1995) Isolation, purification and structure of exochelin MS, the extracellular siderophore from *Mycobacterium smegmatis*. Biochem J 305:187–196

95. Sharman GJ, Williams DH, Ewing DF et al (1995) Determination of the structure of exochelin MN, the extracellular siderophore from *Mycobacterium neoaurum*. Chem Biol 2:553–561

96. Hall RM, Wheeler PR, Ratledge C (1983) Exochelin-mediated iron uptake into *Mycobacterium leprae*. Int J Lepr 51:490–494

97. Goodfellow M, Wayne LG (1982) Taxonomy and nomenclature. In: Ratledge C, Stanford J (eds) The biology of the mycobacteria, vol 1. Academic Press, London

98. Stanford JL, Rook GAW (1976) Taxonomic studies on the leprosy bacillus. Int J Lepr 44:216–221

99. Barclay R, Ratledge C (1988) Mycobactins and exochelins of *Mycobacterium tuberculosis*, *M. bovis*, *M. africanum* and other related species. J Gen Microbiol 134:771–776

100. Macham LP, Ratledge C, Nocton JC (1975) Extracellular iron acquisition by mycobacteria: role of the exochelins and evidence against the participation of mycobactin. Infect Immun 12:1242–1251

101. Lane SJ, Marshall PS, Upton RJ et al (1995) Novel extracellular mycobactins, the carboxymycobactins from *M. avium*. Tetrahedron Lett 36:4129–4132

102. Gobin J, Moore CH, Reeve JR, Wong DK et al (1995) Iron acquisition by *Mycobacterium tuberculosis*: isolation and characterization of a family of iron-binding exochelins. Proc Nat Acad Sci USA 92:5189–5193

103. Wong DK, Gobin J, Horwitz MA, Gibson BW (1996) Characterization of exochelins from *Mycobacterium avium*: evidence for saturated and unsaturated and for acid and ester forms. J Bacteriol 178:6394–6398

104. Lane SJ, Marshall PS, Upton RJ, Ratledge C (1998) Isolation and characterization of carboxymycobactins as the second extracellular siderophores in *Mycobacterium smegmatis*. Biometals 11:13–20

105. Homuth M, Valentin-Weigand P, Rohle M et al (1998) Identification and characterization of a novel extracellular ferric reductase from *Mycobacterium paratuberculosis*. Infect Immun 66:710–716

106. Lambrecht RS, Collins MT (1993) Inability to detect mycobactin in mycobacteria-infected tissues suggests and alternative iron acquisition mechanism by mycobacteria in vivo. Microb Pathog 14:229–238

107. Emmanuel FX, Seagar AL, Doig C et al (2007) Human and animal infections with *Mycobacterium microti*, Scotland. Emerg Infect Dis 13:1924–1927

108. Ratledge C, Ewing M (1996) The occurrence of carboxymycobactin, the siderophore of pathogenic mycobacteria, as a second extracellular siderophore in *Mycobacterium smegmatis*. Microbiology 142:2207–2212

109. Brooks BW, Yong NM, Watson DC et al (1991) *Mycobacterium paratuberculosis* antigen D: characterization and evidence that it is bacterioferritin. J Clin Microbiol 29:1652–1658

110. Inglis NF, Stevenson K, Hosie AH et al (1994) Complete sequence of the gene encoding the bacterioferritin subunit of *Mycobacterium avium* subspecies *silvaticum*. Gene 150:205–206

111. Pessolani MCV, Smith DR, Rivoire B et al (1994) Purification, characterization, gene sequence, and significance of a bacterioferritin from *Mycobacterium leprae*. J Exp Med 180:319–327
112. Matzanke BF, Bohnke R, Mollmann U et al (1997) Iron uptake and intracellular metal transfer in mycobacteria mediated by xenosiderophores. Biometals 10:193–203
113. Ratledge C (1999) Iron metabolism. In: Ratledge C, Dale J (eds) Mycobacteria: molecular biology and virulence, Blackwell, Oxford, pp 260–286

Chapter 3
Mycobacterial Iron Acquisition Mechanisms

B. Rowe Byers

Abstract In both pathogenic and saprophytic mycobacteria, many of the genes and systems required for high affinity iron acquisition have been identified, including siderophore production, uptake of ferric-siderophores, production of iron storage proteins, and uptake of heme. Production and function of iron uptake mechanisms is controlled by a regulatory protein. Possible low affinity acquisition through multiple function porins also has been described. In pathogenic mycobacteria, most of the high affinity systems appear necessary for maintenance of an infection. Greater definition of the functions of both the identified genes and genes yet to be discovered will refine our understanding of mycobacterial iron acquisition and the interplay between components of the iron systems.

Keywords Iron • Siderophores • Mycobactin • Carboxymycobactin • Exochelin • Mycobacterial heme utilization • Ferritin • Bacterioferritin • Continuous culture of mycobacteria

3.1 The Need for Siderophores in Iron Acquisition

Just as iron is essential in the metabolism of almost all modern biological cells, so the metal was crucial in the chemical reactions that created life. Assuming that life did not arrive on Earth by seeding from an extraterrestrial source, it has been proposed that the energy driving the primordial system was provided by the reaction converting ferrous ions and hydrogen sulfide to iron pyrite [1]. This is thought to have occurred in the upwelling heated water at the deep sea juncture of tectonic plates. The non-protein iron sulfur clusters that formed at the edges of

---
B. R. Byers (✉)
The University of Mississippi Medical Center, Jackson, MS, USA
e-mail: rowebyers@comcast.net

these mineral-rich hydrothermal mounds served as early energy transforming molecules for carbon fixation by CO_2 reduction [2, 3]. Current dependency of most energy transducing metabolic sequences on iron containing cofactors argues for pivotal use of the primordial versions of these cofactors in the expansion of these reactions. Subsequent enclosure of the nascent biological system in a membrane vesicle allowed development of a greater range of iron cofactors, including those required for redox reactions, as well as the iron sensors of cellular redox status and ferritin-like proteins for iron storage and protection from iron induced damage [4]. For example, the iron-sulfur complexes (with a later acquired protein component) probably were the precursors of today's ferredoxins [1] and in some microorganisms non-heme iron proteins can replace the coenzyme NADP [5]. The encircling membrane also permitted control of the entry and exit of substances. As early Earth is considered to have been anaerobic, iron would have been present in its soluble ferrous form and transport of the metal into the vesicle could have used the early prototypes of metal symporters or antiporters.

In an unusual comparison to human technology, Wachtershäuser [1] suggested that cell envelopes were like space suits which enabled existence in uninhabitable regions. The surface of the planet was abundant with radiant energy from the sun and within their "space suits" cells could both venture to the surface and evolve to fit the new environment, utilizing light as an energy source not available to subsurface cells [6, 7]. However, most of the required iron-promoted reactions may have been in place before the origin of light capturing pigments completed the route to anoxygenic photosynthesis. Today, some photoferrotrophic bacteria still display a photobiologic process in which ferrous iron is oxidized with concomitant reduction of CO_2 to cell material, implying that anoxygenic photosynthesis was present before oxygenic photosynthesis appeared.

The present increase in the planet's temperature may be due in part to increase in greenhouse gases from human activities. Similarly, about 2.4 billion years ago the anoxygenic life-form underwent a change to oxygenic photosynthesis in which water served as the electron donor and oxygen was evolved [8–10]. Iron components remained as essential units in the controlled downhill flow of electrons from photosynthetic pigments and the repercussions of this change led to development of new biosynthetic pathways, modification of electron carriers, and alternative methods of carbon fixation. The demand for iron may have been amplified by these changes.

The rise in oxygen level linked to oxygenic photosynthesis is a good example of how a life-form can cause a self-inflicted alteration the planet's atmosphere. As oxidation progressed, reduced iron was changed from its water soluble ferrous valence to the nearly water insoluble oxidized ferric ion with resultant precipitation of the metal from solution. Under these oxidized circumstances life could retreat to, or remain in anaerobic niches or regions of low oxygen tension (some did), find substitutes for iron (very few did), or capture ferric iron and deliver it to the cells (most did). Ferric chelating siderophores that extracted the metal from insoluble complexes and delivered the precious commodity to bacteria were a remarkable answer to the iron acquisition problem in oxidized environments. The term siderophore ("iron bearer") is an insightful trivial designation made by Charles Lankford in 1973 [11] which like all useful trivial names for complex molecules conveys information that recalls for the reader the outstanding function, location, or other relevant property of the molecule. An outstanding chemical asset of siderophores is their competitive binding affinity for ferric iron but very low affinity for ferrous iron, allowing the metal to be released from the ferric-siderophore by reduction once the ferric-siderophore is transported internally by siderophore receptors and transport systems. The molecular mass (usually about 500–1000 Da) is too large to move internally across the membrane without a cognate system.

Most siderophores belong to three groups: the catecholate structural group, which is usually based on 2, 3-dihydroxybenzoic acid (2, 3-DHB) components, the hydroxamic acids, and mixed types. Production of siderophores is boosted by decreased environmental iron levels, implying careful cellular monitoring of the internal iron status. Interestingly, the simple subunit 2, 3-DHB is often released by catecholate siderophore producers into the extracellular region. In a unique instance, *Bacillus anthracis* and microorganisms related to *B. anthracis* produce the catecholate siderophore petrobactin which is based on the phenolate 3, 4-dihydroxybenzoic acid (3, 4-DHB) [12]. Like 2, 3-DHB, the 3, 4-DHB subunit of petrobactin is found external to the microbial cells during iron restriction. In the mycobacteria, excretion of salicylic acid, a common moiety of cell associated mycobactin and excreted carboxymycobactin (see Chap. 2), is promoted by iron starvation. It has been proposed that a modern function of salicylic acid might be to act as an internal acceptor of ferrous iron once the metal has been reduced and released from ferric-siderophores [13]. Similarly, the simple catecholate subunits might be internal acceptors of the ferrous ion discharged by reduction from ferric-siderophores. If siderophore subunits have internal ferrous ion acceptor roles, then why is there such a dramatic extracellular accumulation of these compounds? We have argued above that primordial versions later became modern iron cofactors or components of iron acquisition processes. As excreted ferrous iron binding agents, these simple compounds may have encouraged ferrous uptake by way of early porin-like membrane constituents on an anaerobic Earth; however, their chelation capacities became inadequate once oxidized conditions predominated the environment and iron was redistributed to the ferric state. It is possible that from the simple excreted moieties, the complex ferric binding siderophores could have been configured to form ferric chelation centers. In other instances some of the siderophores are built on a backbone of citric acid, inferring that citric acid may be a metal ion transporter that was modified to yield some of the high affinity ferric chelating siderophores.

3.2 The "Iron Biofulcrum" of Infectious Diseases

An invading pathogenic microorganism must interrupt the normal flow of iron that persists in nearly a closed circuit in the vertebrate host. Table 3.1 lists most of the known host and microbial factors involved in this pivotal struggle [14]. While iron acquisition is far from the only trait required for virulence, the pathogen must tilt

Table 3.1 The Iron Fulcrum in Infectious Diseases[a]

Pathogen	Iron	Host
Iron acquisition factors	← Iron →	Iron-withholding defenses
Ferric siderophores (bind iron from host iron containing molecules including transferrin, lactoferrin, ferritin, except heme)		Iron management system; Hypoferremic response; Siderocalin inactivation of siderophores
Ferrous uptake (surface or excreted reductases)		
Transferrin receptors (possible reduction with subsequent ferrous uptake)		
Heme receptors and uptake (digestion of heme-proteins)		Heme and hemoglobin binding proteins (hemopexin, haptoglobin)
Other?		Other?

[a]Modified from Ref. [14]

the iron fulcrum in its favor to gather iron, survive and multiply. The dangerous but necessary reactivity of iron in metabolism means that inadvertent exposure of iron must be avoided. This is accomplished with an iron management process that moves iron throughout the vertebrate host while bound in ferric-transferrin and which also produces protective iron storage molecules that sequester the metal. This iron control process creates an iron-poor circumstance that tends to seclude the metal from the microbial invader. When faced with a microbial infection the host may mount the so-called hypoferremic response in which the iron concentration in transferrin is drastically lessened. Some pathogens use their siderophores to capture iron from the host but recently the host protein siderocalin was shown to be another player in the iron-withholding defense strategies. Siderocalin inactivates some siderophores, forcing productive pathogens to synthesize siderophores resistant to siderocalin. However, successful pathogens have overcome the iron-withholding restrictive barriers. Reviews on the subject of iron and infection [15–21] and Chapter Four of the present volume should be consulted for more information.

3.3 The Mycobacterial Siderophores and Iron Acquisition: Mycobactin, Carboxymycobactin, and Exochelin

An exhaustive review of the literature on iron acquisition in the mycobacteria will not be made; recent pertinent reviews should be consulted for detailed earlier information [18, 22–24], (also see Chap. 2). Evidence from gene expression studies shows that *M. tuberculosis* faces iron restriction during growth in lungs and human macrophages and mutant strains deficient in iron acquisition are attenuated for growth in macrophages [22, 23, 25–28].

All mycobacteria with very few exceptions produce two salicylate containing siderophore subgroups with the same core structure [23]. Mycobactin is a cell-associated molecule, whereas the second salicylate siderophore, called initially carboxymycobactin (see Chap. 2), contains a side chain that confers water solubility and is excreted from the mycobacterial cells. A cluster of genes (designated *mbt*) encodes both mycobactin and carboxymycobactin; *mbt* mutants fail to synthesize the mycobactin core and produce neither mycobactin nor carboxymycobactin [29]. A third siderophore called exochelin is excreted by only the saprophytic mycobacteria (see Chap. 2 for siderophore structures). Biosynthesis of exochelin is encoded by at least three genes: *fxbA* which adds the formyl group to exochelin and *fxbB*, *fxbC* which share homology with motifs common to non-ribosomal peptide synthetases [30–32]. Another gene product ExiT is a member of the superfamily of ABC transporters and may be part of the exochelin export system that delivers exochelin to the exterior of the saprophyte *Mycobacterium smegmatis* [32].

Similar to all siderophores, production of mycobactin, carboxymycobactin, and exochelin is increased by iron restrictive cultivation. The iron dependent regulatory protein IdeR controls production of the mycobactin class of siderophores in *M. tuberculosis* [33]. Consensus "IdeR boxes" were identified in the saprophyte *M. smegmatis* [34], suggesting that *M. smegmatis* also employs IdeR control. In low iron conditions, IdeR relieves repression of siderophore genes, ultimately promoting iron uptake. When sufficient iron is present in the mycobacterial cell, IdeR activates transcription of genes for ferritin and bacterioferritin to sequester iron and protect the mycobacterial cell from iron induced damage [20, 24, 33]. Multiple other genes also are either repressed or derepressed by IdeR regulation [33].

It was suggested that mycobactin could be a temporary iron storage agent [18] and iron exchange from ferric-carboxymycobactin to mycobactin has been noted [35]. As both of the mycobactin class of siderophores have the same binding affinities, exchange should favor iron movement into mycobactin from a greater concentration of ferric-carboxymycobactin obtained by external chelation. Such iron exchange between the two siderophores may reveal only what is plausible, not the major route of metal uptake because *mbt* mutants that make neither of the mycobactin siderophores can use exogenously provided ferric-carboxymycobactin, suggesting that cell-associated mycobactin is non-essential for iron uptake from ferric-carboxymycobactin [20]. Inactivation of two IdeR-regulated genes *irtA* and *irtB* that encode a putative ABC transporter showed these genes to be involved in uptake of ferric-carboxymycobactin in *M. tuberculosis* in macrophages [36, 37]. The gene products IrtAB are not required for carboxymycobactin excretion but are components of the ferric-carboxymycobactin uptake mechanism. Using reconstituted proteoliposomes, it was argued by others that IrtA exported carboxymycobactin and IrtB imported ferric-carboxymycobactin [38]; however, other work indicated that both IrtAB were required for uptake of ferric-carboxymycobactin [37]. Despite the blockade in *M. tuberculosis* high affinity uptake of ferric-carboxymycobactin caused by mutations in the *irtA* and *irtB* genes, iron acquisition was not completely eliminated and some iron was accumulated. Upon inactivation of *irtAB*, another channel may have been opened possibly through iron transferred to cell-associated mycobactin from ferric-carboxymycobactin. Because many bacterial ferric reductases are flavin adenine dinucleotide (FAD) reductases and IrtA binds a molecule of FAD, it could be a reductase of ferric-carboxymycobactin imported by IrtAB similar to the cytosolic ViuB protein required for utilization of the siderophore ferric-vibriobactin by *Vibrio cholerae*. In *M. tuberculosis* the locus Rv2895c, a possible homolog of ViuB, was postulated to be involved in iron acquisition but contrary to this prediction, analysis indicates that Rv2895c was not required for iron uptake in *M. tuberculosis* [39].

Earlier research clarified, in part, the genes involved in uptake of ferric-exochelin in *M. smegmatis* [30]. Uptake is accomplished by the products of four genes: *fxuA*, *fxuB*, *fxuC*, and *fxuD*.

The mycobacterial cell wall may impede ready exchange of metabolites in terrestrial microorganisms and in pathogenic mycobacteria during an infection. The genetic locus *esx-3* encodes an IdeR regulated secretion process required for acquisition of iron from the mycobactin siderophores [27]. Unlike pathogenic mycobacteria which make only cell bound mycobactin and excreted carboxymycobactin, the exochelin-producing saprophyte *M. smegmatis* tolerated deletion of *esx-3*. Using strains of *M. smegmatis* that lack *esx-3* with combinations of deficiencies in the mycobactin/carboxymycobactin biosynthetic pathway and in biosynthesis of exochelin, the potential interaction of Esx-3 and siderophore production in iron uptake was assessed. An *M. smegmatis* mutant strain unable to produce both the mycobactins and exochelin siderophores was rescued by addition of exogenous mycobactin or carboxymycobactin; if Esx-3 was missing neither mycobactin nor carboxymycobactin rescued this strain. *M. smegmatis* lacking both Esx-3 and exochelin production, but still able to produce mycobactin/carboxymycobactin, had a severe low iron growth defect that was complemented by Esx-3. Therefore, Esx-3 was required for mycobactin utilization and would be essential for growth in the host. In *M. smegmatis*, production of exochelin appeared to by-pass the use of mycobactin; a role for Esx-3 was evident only in strains lacking exochelin production in which Esx-3 was required for use of mycobactin as a possible back-up process in iron uptake. The function of *esx-3* is uncertain, although it may be involved in correct positioning of components in the cell wall or at the cell surface.

The saprophyte *M. smegmatis* has an ABC-type transporter ExiT that is responsible for export of exochelin [32]. Given the external location of carboxymycobactin, it is probable that an export system also exists in *M. tuberculosis* for this siderophore. Recently, two iron (IdeR) regulated genes *mmpS4* and *mmpS5* that encode two outer membrane proteins were identified [40]. Results obtained with deletion mutants indicated that neither of the membrane proteins MmpS4 MmpS5 was involved in uptake of ferric-carboxymycobactin. Deletion of both genes created a mutant strain with a growth defect in low iron culture medium. The double mutant failed to proliferate in lungs and spleen of infected mice but the mutant also made significantly less carboxymycobactin and mycobactin than the wild-type parental culture. It was suggested that the lowered levels of the mycobactin siderophores in the double deletion mutant was due to decreased production of these siderophores because the mycobactin core biosynthetic enzyme MbtG may be inner membrane associated [40]. It was postulated that *mmpS4* and *mmpS5* are genes encoding components of a siderophore export system essential for virulence of *M. tuberculosis* [40] and the inability to export carboxymycobactin might lead to inefficient siderophore synthesis, coupling synthesis with secretion. The alveolar macrophage in which *M. tuberculosis* grows maintains a low level of phagosomal iron yet the wild-type pathogen overcomes this iron-withholding defense probably using its mycobactin class of siderophores. It is uncertain if the lowered amounts of siderophore produced by the *mmpS4* and *mmpS5* double deletion mutant altered iron acquisition and thereby affected virulence.

3.4 Mycobacterial Iron Acquisition from Heme

Similar to many other pathogenic microorganisms, the mycobacteria have not ignored the 70 % of total iron in a mammal that is present in heme [41–43]. The growth defect noted in an *M. tuberculosis* strain deficient in production of the mycobactin class of siderophores could be satisfied with heme [42]. It is uncertain how heme is taken up by the mycobacteria and heme must be degraded for redistribution of iron into various metabolic systems. Lysins and proteases may be required to make heme available. Contact dependent hemolytic activity has been reported in *Mycobacterium avium* and *M. tuberculosis* [43, 44].

Whereas in liquid medium, heme restored growth of a double deletion mutant of the *mmpS4* and *mmpS5* genes (described above in Sect. 3.4) to the wild-type level, growth of the double mutant on agar was not fully restored by addition of hemoglobin to the agar [40]. Interestingly, this growth impairment of the double mutant was abolished if the double mutant also could not synthesize the carboxymycobactin/mycobactin siderophore group, illustrating one of the complexities of iron acquisition from sources other than the siderophores.

3.5 Ferritin and Mycobacterial Persistence in Animals

Because of the capacity of iron to catalyze production of free radicals, it is essential the mycobacteria sequester intracellular iron yet make the metal readily available for insertion into metabolism. Most of the intracellular iron may be present in iron storage proteins [46] and two iron storage proteins have been identified, namely BfrA (a bacterioferritin) and BfrB (a ferritin-like protein) [47]. Disruption of the relevant genes *bfrA* and *bfrB* in *M. tuberculosis* significantly lowered in vitro growth of the microorganism and attenuated growth in human macrophages [48]. The expression of both iron storage genes is regulated by the iron control protein IdeR. A mutant strain of *M. tuberculosis* lacking both *bfrA* and *bfrB* also is markedly sensitive to oxidative stress agents, demonstrating the need for iron storage proteins in prevention of iron induced damage.

Individual knockout of *M. tuberculosis bfrA* an *bfrB* revealed that ferritin may be essential for the maintenance of iron homeostasis [49]. *M. tuberculosis* lacking ferritin suffered from iron induced toxicity, was unable to persist in mice and was highly susceptible to killing by antibiotics. The iron storage proteins are essential components of the iron acquisition process.

3.6 Low Affinity and Reductive Mycobacterial Iron Acquisition

The saprophytic *M. smegmatis* exhibits low affinity iron capture from sources like ferric citrate if present at elevated levels of iron [50]. The Msp porins of *M. smegmatis* are likely responsible for low affinity uptake because ferric-exochelin, the siderophore of saprophytes, is acquired by a porin independent process involving the *fxuA-D* genes. At high iron concentrations, the requirement for iron may be met by low affinity uptake through porins. When the iron level falls or the number of porins is lessened, *M. smegmatis* derepresses production of siderophores for high affinity iron gathering. Where in the terrestrial environment a saprophyte would find elevated iron is uncertain. It also seems unlikely that a strict pathogen like *M. tuberculosis*, which moves from the iron restricted condition of a human host to another susceptible human, would encounter iron replete circumstances. Similarly, an opportunistic mycobacterium causing an infection would face host iron withholding.

It is generally assumed that siderophores are deferrated internally by reduction of the metal; however, bacterial cell surface reductases may be required to render iron to its soluble state for transport in some bacteria. Microorganisms like *Streptococcus mutans* which does not produce siderophores likely require surface reduction of iron prior to transport [51]. A novel extracellular ferric reductase was discovered in *Mycobacterium paratuberculosis* capable of scavenging iron from ferric-transferrin, ferritin, and ferric citrate [52]. Reduction of iron at the cell surface is an alternative to iron capture by high affinity siderophores and may be a major route in a few mycobacteria.

3.7 Iron and Continuous Growth of *M. smegmatis* in a Teflon Chemostat: Biofilm Formation

Continuous cultivation of *M. smegmatis* was accomplished in a chemostat constructed of Teflon which overcomes the problem of metal leaching from glass and stainless steel components [52, 53]. There are few reports of continuous growth of mycobacteria [54]. Unpublished studies [B. R. Byers and J. E. L. Arceneaux] using a defined medium supplemented with high purity metals found that steady-state growth of *M. smegmatis* required an iron addition of 0.4–0.8 μM. Higher levels of iron did not significantly increase steady-state growth but did cause changes in protein expression, indicating a response to the iron level in the chemostat. Although Tween 80 addition was not required for batch cultivation of *M. smegmatis*, Tween 80 was required for establishment of steady-state growth. Flow cytometry with fluorocein diacetate and propidium iodide staining revealed that most of the mycobacterial cells in the chemostat were viable and numbers obtained by plate counting were in good correlation. Using organisms like *M. avium* and *M. tuberculosis*, steady-state growth studies at different iron levels will allow examination of various parameters, including siderophore gene expression, protein expression, iron uptake kinetics, and the capacity to infect host cells.

Many mycobacteria form biofilms at liquid–air interfaces and on solid surfaces [55]. Development of *M. smegmatis* biofilms required an iron supplement and was accompanied by induction of siderophore genes. While it is difficult to assess the effects of iron on biofilm growth because different concentrations of iron probably exist within a biofilm leading to differences in siderophore biosynthesis, iron levels influence mycobacterial biofilm formation.

3.8 The Complexity and Perplexity of Mycobacterial Iron Acquisition: Conclusions

The complexity of mycobacterial iron uptake is illustrated by the several routes for iron procurement found in these microorganisms. Multiple ways may exist to pursue the single purpose of iron capture, an apparent redundancy that causes confusion in identification of which of the iron uptake tracks looms critical under the experimental conditions being tested. A perplexing question is whether there is an interaction between mycobactin and carboxymycobactin of pathogens and between the mycobactin siderophores and exochelin of the terrestrial saprophytes. The answer to this question may be both yes and no and will vary with test conditions. Adventitious iron in the culture system, prior growth conditions, inoculum size, and the amount of iron supplied to the microorganism as well as the species tested and other undefined conditions will influence results and must be considered. Iron acquisition paths may overlap, intersect, or be a single conduit for the flow of iron into metabolism. If excreted carboxymycobactin can replace exochelin in mutants unable to produce exochelin, then why is exochelin necessary? Exochelin may represent the most direct path for iron to reach metabolism in a saprophyte. While iron uptake through the carboxymycobactin/mycobactin system of pathogenic mycobacteria may appear to be a complex route, the mycobactin siderophores may confer special capacities needed for growth of mycobacteria in host tissue and macrophages. If possible, it might be helpful to ascertain the iron acquisition capacities of mutants altered in production of either mycobactin or carboxymycobactin [24]. This probably would require alteration of the side chains attached to the mycobactin core structure. A mutation in the *M. smegmatis* ortholog of *fad33* disrupted mycobactin synthesis resulting in a molecule with an altered side chain that may reveal an intersection between iron acquisition and lipid metabolism [34]. The product of *fad33* may be an acyl-coenzyme A synthase and mutations in this gene could produce alterations in mycobactin side chains with resultant changes in iron uptake through the mycobactin system.

As described above, many of the genes and systems required for high affinity mycobacterial iron acquisition have been identified, including siderophore production, siderophore export, uptake of ferric-siderophores, assembly of iron storage proteins, uptake of heme, and regulation of these processes, as well as possible low affinity acquisition through multiple function porins and iron transport subsequent to its reduction. In pathogenic *M. tuberculosis*, most of the high affinity systems appear necessary for maintenance of infection, indicating a role in iron acquisition in the macrophage. Better definition of the functions of the identified genes, as well as genes yet to be discovered (including a possible outer receptor for ferric-carboxymycobactin), will refine our understanding of mycobacterial iron acquisition. Schematic drawings illustrating the complex interplay of the iron uptake processes were not attempted here. Reference is made to Fig. 11 of Chap. 2 of this volume which depicts the mycobactin and exochelin systems and which outlines the platform on which many of the present genetic studies were based.

References

1. Wachtershäuser G (1988) Before enzymes and templates: theory of surface metabolism. Microbiol Rev 52:452–484
2. Koch AL, Schmidt TM (1991) The first cellular bioenergetic process: primitive generation of a proton-motive force. J Mol Evol 33:297–304
3. Woese CR (1979) A proposal concerning the origin of life on planet earth. J Mol Evol 13:95–101
4. Outten FW, Theil EC (2009) Iron-based redox switches in biology. Antiox Redox Signal 11:1029–1046
5. Daniel RM, Danson MJ (1995) Did primitive microorganisms use nonhem iron proteins in place of NAD/P? J Mol Evol 40:559–563
6. Saraiva IH, Newman SK et al (2012) Functional characterization of the FoxE iron oxidoreductase from the photoferrotroth Rhodobacter ferrooxidans SW2. J Biol Chem 287:25541–25548
7. Widdel F, Schnell S et al (1993) Ferrous iron oxidation by anoxygenic phototrophic bacteria. Nature 362:834–836
8. Grenfell JL, Rauer H et al (2010) Co-evolution of atmospheres, life, and climate. Astrobiol 10:77–88
9. Hohmann-Marriot MF, Blankenship RD (2011) Evolution of photosynthesis. Annu Rev Plant Biol 62:515–548
10. Trevors JT (2002) The subsurface origin of microbial life on Earth. Res Microbiol 153:487–491
11. Lankford CE (1973) Bacterial assimilation of iron. Crit Rev Microbiol 2:273–331
12. Garner BL, Arceneaux JEL et al (2004) Temperature control of a 3, 4-dihydroxybenzoate (protocatechuate)-based siderophore in *Bacillus anthracis*. Curr Microbiol 49:89–94
13. Adilakshmi T, Ayling PD et al (2000) Mutational analysis of a role for salicylic acid in iron metabolism of *Mycobacterium smegmatis*. J Bacteriol 182:264–271

14. Byers BR, Arceneaux JEL (1998) Micobial iron transport: iron acquisition by pathogenic microorganisms. In: Sigel A, Sigel H (eds) Metal ions in biological systems, vol 35 Iron Transport and Storage in Microorganisms, Plants and Animals. Marcel Dekker, New York
15. Bullen JJ, Griffiths E (eds) (1999) Iron and infection: molecular, physiological and clinical aspects, 2nd edn. Wiley, Chichester
16. Cornelis P, Andrews SC (eds) (2010) Iron uptake and homeostasis in microorganisms. Caister Academic Press, Norfolk
17. Nairz N, Schroll A et al (2010) The struggle for iron-a metal at the host-pathogen interface. Cell Microbiol 12:1692–1702
18. Ratledge C (2004) Iron, mycobacteria and tuberculosis. Tuberculosis 84:110–130
19. Ratledge C, Dover G (2000) Iron metabolism in pathogenic bacteria. Annu Rev Microbiol 54:881–941
20. Rodriguez M, Smith I (2003) Mechanisms of iron regulation in mycobacteria: role in physiology and virulence. Molec Microbiol 47:1485–1494
21. Saha R, Saha N et al (2012) Microbial siderophores: a mini review. J Basic Microbiol 52:1–15
22. De Voss JJ, Rutter K et al (1999) Iron acquisition and metabolism by mycobacteria. J Bacteriol 181:4443–4451
23. De Voss JJ, Rutter K et al (2000) The salicylate-derived mycobactin siderophores of *Mycobacterium tuberculosis* are essential for growth in macrophages. Proc Natl Acad Sci USA 97:1252–1257
24. Rodriguez GM (2006) Control of iron metabolism in *Mycobacterium tuberculosis*. Trends in Microbiol 14:320–327
25. Gold B, Rodriguez GM et al (2001) The *Mycobacterium tuberculosis* IdeR is a dual function regulator that controls transcription of genes involved in iron acquisition, iron storage, and survival in macrophages. Mol Micro 42:851–865
26. Schnappinger E, Vaskuil MI et al (2003) Transcriptional adaptation of *Mycobacterium tuberculosis* within macrophages: insights into the phagosome environment. J Exp Med 198:693–704
27. Siegrist MS, Unnikrishnan M et al (2009) Mycobacterial Esx-3 is required for mycobactin-mediated iron acquisition. Proc Natl Acad Sci USA 106:18792–18797
28. Timm J, Post FA et al (2003) Differential expression of iron-, carbon- and oxygen-responsive mycobacterial genes in the lungs of chronically infected mice and tuberculosis patients. Proc Natl Acad Sci USA 100:14321–14326
29. McMahon MD, Rush J et al (2012) Analyses of MbtB, MbtE, and MbtF suggest revisions to the mycobactin biotynthesis pathway in *Mycobacterium tuberculosis*. J Bacteriol 194:2809–2818
30. Fiss EH, Yu S, Jacobs WR Jr (1994) Identification of genes involved in sequestration of iron in mycobacteria: the ferric exochelin biosynthetic and uptake pathways. Mol Microbio 14:557–569
31. Yu S, Fiss E, Jacobs WR Jr (1998) Analysis of the exochelin locus in *Mycobacterium smegmatis*: biosynsthesis genes have homology with genes of the peptide synthetase family. J Bacteriol 180:4676–4685
32. Zhu W, Arceneaux JEL et al (1998) Exochelin genes in *Mycobacterium smegmatis*: identification of an ABC transporter and two non-ribosomal peptide synthetase genes. Mol Micobiol 29:629–639
33. Rodriguez GM, Voskuil MI et al (2002) ideR, an essential gene in *Mycobacterium tuberculosis*: role of ideR in iron-dependent gene expression, iron metabolism, and oxidative stress response. Infect Immun 70:3371–3381
34. LaMarca BBD, Zhu W et al (2004) Participation of fad and mbt genes in synthesis of mycobactin in *Mycobacterium smegmatis*. J Bacteriol 186:374–382
35. Gobin J, Horwitz MA (1996) Exochelins of *Mycobacterium tuberculosis* remove iron from human iron-binding proteins and donate iron to mycobactins in the *Mycobacterium tuberculosis* cell wall. J Expt Med 183:1527–1532
36. Rodriguez GM, Smith I (2006) Identification of an ABC transporter required for iron acquisition and virulence. J Bacteriol 188:424–430

37. Ryndak MB, Wang S et al (2010) The *Mycobacterium tuberculosis* high affinity iron importer, IrtA, contains an FAD-binding domain. J Bacteriol 192:861–869
38. Faranh A, Kumar S et al (2008) Mechnistic insights into a novel exporter-importer system of *Mycobacterium tuberculosis* unravel its role in trafficking of iron. PLoS One 3:e2087
39. Santhanagipaken S, Rodgriguez GM (2011) Examining the role of Rv2895c (ViuB) in iron acquisition in *Mycobacterium tuberculosis*. Tuberculosis 92:60–62
40. Wells RM, Jones CM, Xi Z et al (2013) Discovery of a siderophore export system essential for virulence of *Mycobacterium tuberculosis*. PLoS Pathog 9(1–14):e1003120
41. Chim N, Iniquez P et al (2010) Unusual diheme conformation of the heme-degrading protein from *Mycobacterium tuberculosis*. J Mol Biol 395:595–608
42. Jones CM, Neiderweis M (2011) *Mycobacterium tuberculosis* can utilize heme as an iron source. J Bacteriol 193:1767–1770
43. Owens CP, Du J et al (2012) Characterization of heme ligation properties of Rv0203, a secreted heme binding protein involved in *Mycobacterium tuberculosis* heme uptake. Biochemistry 51:1518–1531
44. Deshpande RG, Khan MB et al (1997) Isolation of a contact-dependent hemolysin from *Mycobacterium tuberculosis*. M Med Microbiol 46:233–238
45. Rindi L, Lari BN et al (2003) Most human isolates of *Mycobacterium avium* Mav-A and Mav-B are strong producers of hemolysin, a putative virulence factor. J Clin Microiol 41:5738–5740
46. Matzanke BF, Böhnke R et al (1997) Iron uptake and intacellular metal transfer in mycobacteria mediated by xenosiderophores. Biometals 10:193–203
47. Reddy PV, Purl RV et al (2012) Iron storage proteins are essential for the survival and pathogenesis of *Mycobacterium tuberculosis* in THP-1 macrophages and the guinea pig model of infection. J Bacteriol 194:567–575
48. Pandey R, Rodriguez GM (2012) A ferritin mutant of *Mycobacterium tuberculosis* is highly susceptible to killing by antibiotics and is unable to establish a chronic infection in mice. Infec Immun 80:3650–3659
49. Jones CM, Neiderweis M (2010) Role of porins in iron uptake by *Mycobacterium smegmatis*. J Bacteriol 192:6411–6417
50. Evans SL, Arceneaux JEL et al (1986) Ferrous iron transport in *Streptococcus mutans*. J Bacteriol 168:1096–1099
51. Homuth M, Valentin-Weigand P et al (1998) Identification and characterization of a novel extracellular ferric reductase from *Mycobacterium paratuberculosis*. Infect Immun 66:710–716
52. Aranha H, Evans SL et al (1982) Effect of trace metals on growth of *Streptococcus mutans* in a teflon chemostat. Infect Immun 35:456–460
53. Strachan RC, Aranha H et al (1982) Teflon chemostat for studies of trace metal metabolism in *Streptococcus mutans* and other bacteria. Appl Environ Microbiol 43:257–260
54. McCarthy CM (1983) Continuous culture of *Mycobacterium avium* limited for ammonia. Amer Rev Respir Dis 127:193–197
55. Ojha A, Hatful GF (2007) The role of iron in *Mycobacterium smegmatis* biofilm formation: the exochelin siderophore is essential in limiting iron conditions for biofilm formation but not for planktonic growth. Mol Microbiol 66:468–483

Chapter 4
Siderocalin Combats Mycobacterial Infections

Benjamin E. Allred, Allyson K. Sia and Kenneth N. Raymond

Abstract The human immunoprotein siderocalin (Scn) protects against infections by binding the siderophores used by a pathogen to steal iron from host iron stores, thereby limiting bacterial growth and subsequent colonization by restricting the supply of iron. *Mycobacterium tuberculosis* synthesizes two siderophores, mycobactin and carboxymycobactin, with the same core structure. Unlike the alkyl side chain of mycobactin, the carboxylate chain of carboxymycobactin imparts water solubility on this siderophore, allowing it to be excreted. The variable length of the carboxymycobactin chain markedly influences binding by Scn. The structural differences in the side chains of mycobactin and carboxymycobactin and the intracellular location of *M. tuberculosis* challenge the ability of Scn to recognize and inactivate siderophore-mediated iron acquisition by this pathogen. We describe the physical interactions between Scn and carboxymycobactin as well as the effect of Scn on the growth of *M. tuberculosis* in different cellular environments. The evidence suggests that Scn is part of the defenses that prevent or limit infections of *M. tuberculosis*.

Keywords Siderocalin • Carboxymycobactin • Mycobactin • Carboxymyc obactin variable side chain length • Carboxymycobactin-siderocalin binding • Siderocalin immune defense against mycobacterial infections

B. E. Allred · A. K. Sia · K. N. Raymond (✉)
University of California, Berkeley, CA, USA
e-mail: raymond@socrates.berkeley.edu

B. E. Allred
e-mail: ballred7@berkeley.edu

A. K. Sia
e-mail: allysonksia@berkeley.edu

B. R. Byers (ed.), *Iron Acquisition by the Genus Mycobacterium,*
SpringerBriefs in Biometals, DOI: 10.1007/978-3-319-00303-0_4,
© The Author(s) 2013

4.1 Siderocalin

The lipocalin family of extracellular proteins has functions that vary significantly among the proteins within this family. From promoting organogenesis and delivering hydrophobic ligands in cell signaling to facilitating biosynthesis and participating in immunodefense, the lipocalins serve both eukaryotic and prokaryotic organisms [1–3]. Despite the diversity of lipocalin functions, these proteins share up to three structurally conserved regions. A characteristic feature of all lipocalins is a barrel formed by eight-stranded anti-parallel beta sheets (Fig. 4.1). The interior binding sites of lipocalins are generally lined with positively charged amino acid residues which create binding interactions with a substrate.

The lipocalin now called siderocalin (Scn) has been known for several decades and designated variously as 24p3, uterocalin, lipocalin 2 (Lcn2), human neutrophil lipocalin (HNL) and neutrophil gelatinase associated lipocalin (NGAL). However, a specific function for it was not fully elucidated until Goetz et al. [4] provided solid-state and solution structural evidence that bacteria-derived ferric enterobactin was complementarily bound to the Scn binding pocket (Fig. 4.1). The most important result from these studies and others which soon followed was that Scn participates in the acute immune response to bacterial infections by sequestering certain ferric siderophores and thus restricting iron piracy and subsequent proliferation by bacteria [4, 5]. For example, the growth of mycobacterial strains in vitro in murine macrophage cell lines was inhibited by addition of recombinant Scn [6]. Scn is also successful in vivo, decreasing the susceptibility of murine models to both *Escherichia coli* and *Mycobacterium tuberculosis* [5, 6]. Other experiments show that Scn-knockout mice have increased susceptibility to bacterial infections

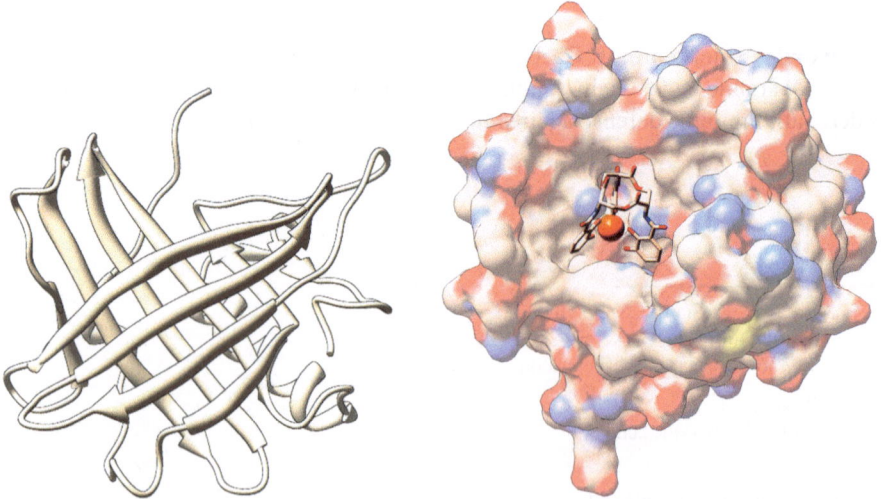

Fig. 4.1 Ribbon-drawing diagram of Scn (*left*) and schematic drawing of Scn with bound ferric-enterobactin (*right*)

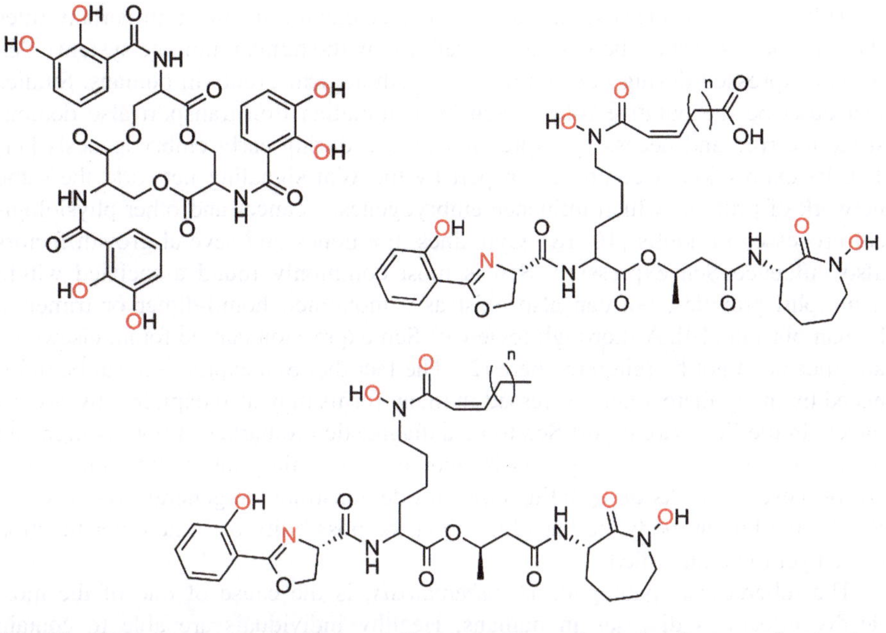

Fig. 4.2 Siderophores enterobactin (*top left*) and generic versions of carboxymycobactins (*top right*, n = 3–10) and mycobactins (*bottom*, n = 6–17). Both carboxymycobactins and mycobactins have varying alkyl substituents on the backbone that, for simplicity, are not included in this figure

and display compromised survival [5, 7]. The role of Scn as an antimicrobial protein was further demonstrated by the compatibility of the Scn calyx with a collection of siderophores from both Gram-positive and Gram-negative bacteria [8–10].

Recent work has also shown that Scn is a mediator of mammalian iron transport, utilizing so-called mammalian siderophores (simple catechols found endogenously in the mammalian gut) [11, 12]. Scn-mediated iron transport is thought to occur in kidney embryogenesis or in cases of kidney damage where concentrations of iron must be strictly regulated to control inflammation. Catechols are iron-binding moieties found in some natural siderophores (e.g., enterobactin, Fig. 4.2) and can be bound by Scn as either the free ligand or the iron complex. Endogenous catechols, found as byproducts of either bacterial or human metabolism, are bidentate iron chelators. Under physiological conditions, ferric *bis*catechol complexes primarily are formed, as determined by speciation calculations [11]. Scn intercepts these complexes and recruits a third catechol to fill the iron coordination shell such that the ferric complex is hexacoordinate. Ferric *tris*catechol complexes carry a -3 charge and the additional aromatic catechol optimizes binding by Scn via hybrid Coulombic and cation-pi interactions. However, not all catechols are mammalian siderophores. Those with substituents on the aromatic ring exhibit a severely compromised affinity for Scn. The consequent decrease in affinity is due to steric clashes between catechol substituents and the rigid Scn calyx [13, 14].

Although the bacteriostatic role of Scn is one of its most important roles (because Scn is part of the first line of defense by the human immune system), Scn is also expressed during both normal and pathological events in humans. Studies that describe the putative role of Scn in mammalian iron transport also demonstrate the role and necessary expression of Scn during early embryogenesis [11, 15]. Its expression is controlled in part by the Wnt signaling network, the same network of proteins which influence embryogenesis, cancer and other physiological processes in adults [16–18]. Cytokines, hormones and several growth factors also influence Scn expression. Scn is most commonly found associated within neutrophil granules, but can also exist as a monomer, homo-dimer or trimer in human plasma [19]. A thorough review of Scn expression can be found elsewhere and thus need not be reiterated here [2]. The fact that Scn expression can be influenced by many factors and expressed in many forms may also explain why several papers in the literature report Scn to be a diagnostic biomarker of both benign and malignant cases, such as in psoriasis, anemia, renal injury and tumorigenesis [2]. All of these networks of signaling cascades demonstrate the general expression of Scn by the human body, perhaps hinting at the possibility of some other function of Scn yet to be identified.

The tuberculosis pathogen, *M. tuberculosis*, is the cause of one of the most deadly infectious diseases in humans. Healthy individuals are able to contain the bacterium in a latent state, thus preventing a *M. tuberculosis* infection from progressing to a disease. This is likely due to several components of the immune system, including Scn, acting in concert to protect the human host. *M. tuberculosis* commonly targets the mammalian respiratory system via alveolar epithelial cells, where the bacterium can reside, replicate and cause a variety of symptoms, including chest pain, fever, and weakness. Furthermore, the bacterium may spread through the body to other systems such as the gastrointestinal or skeletal systems, compromising a patient's general health. The tuberculosis pathogen can be deadly for immunocompromised patients, such as those with human immunodeficiency virus or those living in poverty without access to proper treatment. Furthermore, various drug-resistant strains of *M. tuberculosis* have evolved as a result of incomplete or ineffective drug treatments, forcing us to search for additional therapeutic measures to fight the bacterium.

4.2 Role of Siderophores in Mycobacterial Infections

Pathogenic mycobacteria encounter an iron limited environment when invading a human host. Host iron concentrations are more than sufficient to support life, but the iron-binding proteins transferrin, lactoferrin, and ferritin regulate and withhold the vital resource. Accessing the host iron is critical to the survival of the pathogen. Mycobacteria obtain host iron by producing small-molecule iron chelators known as siderophores that deliver iron to the pathogen. The ability to produce siderophores directly contributes to mycobacterial virulence [20].

To acquire host iron, pathogenic mycobacteria use two structurally-related families of siderophores: mycobactins and carboxymycobactins (Fig. 4.2) [21]. Both families use one hydroxyphenyl-oxazoline and two hydroxamate groups to coordinate iron. The iron binding units are linked by a lysine derived backbone with various alkyl substituents. The mycobactins have one hydroxamate attached to a long saturated or unsaturated alkyl chain (n = 6–17), making the siderophores lipophilic [22–24]. The carboxymycobactins have the same hydroxamate attached to a shorter saturated alkyl chain (n = 3–10) that terminates in a carboxylate, making the siderophores hydrophilic [10, 22, 25]. The opposite solubility of carboxymycobactins and mycobactins is a distinguishing feature between the two siderophore families.

Synthesis of both carboxymycobactin and mycobactin is a response to iron-limited conditions and the structural similarity of the two alludes to an analogous biosynthesis. Both siderophore families are products of the same non-ribosomal peptide synthetase encoded by the gene loci *mbt-1* and *mbt-2* [26–28]. Each gene loci contains binding sites for IdeR, an iron-responsive DNA-binding protein. IdeR binds iron in iron-replete conditions and represses siderophore biosynthesis. When iron levels fall, apo-IdeR releases DNA and siderophore synthesis is derepressed [26, 29].

Even though the structure and regulation of carboxymycobactin and mycobactin are directly related, the mechanism of iron acquisition is dissimilar due to the differing solubilities of each siderophore. Hydrophilic carboxymycobactin is secreted to the extracellular environment to bind host iron and transport it to the pathogen. This mechanism has been observed in most bacterial siderophores; however, the extracellular environment of mycobacteria is unique because the pathogen inhabits macrophage phagocytes. To grow in the iron-limiting conditions found in macrophages, *M. tuberculosis* needs carboxymycobactin to steal iron from the host [29, 30]. Human transferrin, lactoferrin, and ferritin, the major iron storage protein of mononuclear phagocytes, all release iron to carboxymycobactin [31]. Both exogenous and endogenous iron sources supply *M. tuberculosis* with iron, although it is not known if carboxymycobactins are confined to the mycobacterial phagosome or if they go to the macrophage cytoplasm and beyond to chelate iron [32, 33]. Carboxymycobactin may not need to travel out of the phagosome because transferrin endocytosed by the macrophage is trafficked to the phagosome [34, 35].

Following iron chelation, the ferric carboxymycobactin complex transports iron to the mycobacterial cytoplasm by two different mechanisms. The first depends on an ABC transport system. The transmembrane proteins IrtA and IrtB transport ferric carboxymycobactin using energy from ATP hydrolysis. These transporters are required for using ferric carboxymycobactin and for replication of *M. tuberculosis* in media, macrophages, and mice [36].

The other transport model indicates that ferric carboxymycobactin may pass through the cell wall through a porin to the periplasm where the lipophilic mycobactins are localized in the cell membrane. The mycobactins receive the iron from ferric carboxymycobactin and passively carry it across the cell membrane [21, 37, 38].

This process depends on carboxymycobactin; without it, mycobacteria-bound myco-bactin does not accumulate iron from extracellular ferric transferrin [31].

In addition to accepting iron from ferric carboxymycobactin, mycobactins also remove iron directly from the host macrophages in a unique mechanism. The lipo-philic siderophores are water soluble enough to diffuse throughout a macrophage while also being membrane permeable. The macrophage internal iron stores are all vulnerable to sequestration by mycobactins. Once iron is bound, the ferric myco-bactins accumulate in lipid droplets of the macrophages to be trafficked to phago-somes where they may deliver the iron to inhabiting mycobacteria [39].

4.3 Scn Binds Carboxymycobactins

Iron piracy by mycobacteria does not go unchallenged by the human immune sys-tem. Scn, being part of the immunoresponse to bacterial infections, attempts to ward off mycobacterial infections by intercepting mycobacterial siderophores.

The Scn ligand binding domain, or calyx, is shallow, broad, and lined with polar and positively charged residues (Arg81, Lys125, Lys134) [4, 40]. It is also quite rigid, with three binding pockets inside the calyx that impose a steric limita-tion on which ligands are Scn-compatible. Furthermore, Scn resists any confor-mational change when exposed to changes in pH, ionic strength or upon ligand binding [4]. The isolation of ferric enterobactin in the protein calyx unveiled the most direct clue behind the function of Scn. Ferric enterobactin is a metal complex carrying a -3 charge due to its three catecholate units. The catecholate moieties are also aromatic and planar, allowing each to fit complementarily into each of the three binding pockets of Scn, optimized by hybrid electrostatic and cation-pi interactions that occur between the positive residues and catecholates as well as the overall negative charge of the complex [14, 41]. The affinity of Scn for ferric enterobactin is so great that the dissociation constant (K_D) for the protein–ligand complex, 0.41 nM [4], rivals the affinity of ferric enterobactin for its innate recep-tor FepA ($K_D = 0.27$ nM) [42] (Table 4.1).

Table 4.1 Dissociation constants (K_D) of various ligands with siderocalin[a]

Ferric (siderophore[b])	K_D (nM)
Fe(Ent)	0.41 (1)
Fe(CMB), n = 3	>9,000
Fe(CMB), n = 4	2,360 (9)
Fe(CMB), n = 5	1,100 (30)
Fe(CMB), n = 6	654 (20)
Fe(CMB), n = 7	128 (1)
Fe(CMB), n = 8	280

[a]From Ref. [10]
[b]*Ent* Enterobactin, *CMB* Carboxymycobactin

Although the Scn calyx seems optimized for binding three aromatic rings in the three binding pockets lined with positive residues, the protein can also accommodate siderophores and iron-siderophore complexes of varying structures, so long as steric and electrostatic/cation-pi requirements are met. So-called stealth siderophores are structurally modified such that these and the ferric complexes preclude binding by Scn due to clashes with Scn binding pocket walls. Examples are the anthrax pathogen siderophore, petrobactin, which has 3,4-catecholate iron binding moieties instead of 2,3-catecholates, and salmochelins, which are glycosylated enterobactin structures. Furthermore, Scn does not usually bind hydroxamate-based siderophores or corresponding iron complexes since these substrates lack the aromatic electronic structure necessary for cation-pi interactions with the protein [14]. On the contrary, siderophores with structures that vary drastically from enterobactin can still be accommodated by Scn. The carboxymycobactins are one example; although these siderophores bind iron with two hydroxamate units and one hydroxyphenyl-oxazoline moiety, some ferric carboxymycobactins are sequestered by Scn, as will be further explained.

Aside from the aforementioned iron-binding moieties, carboxymycobactins also have a fatty acid tail of varying lengths (Fig. 4.2, n = 3–10) which has some impact on Scn affinity for the corresponding ferric complexes [10]. The affinities of ferric carboxymycobactin structures for Scn have been evaluated in vitro with fluorescence quenching binding assays to determine the protein–ligand dissociation constants (K_D) as well as with protein crystallography to obtain an accurate, three-dimensional input as to why certain isoforms are better ligands than others. Hoette et al. [10] (Table 4.1) have shown that Scn has the greatest affinity for carboxymycobactin isoform of n = 7, 8 and cannot sequester isoforms as well for n = 3–6. The K_D for the n = 7 carboxymycobactin isoform was the lowest (128 nM), while some others had much higher K_D's (e.g. >900 nM for n = 3) essentially indicating no significant affinity.

Crystal structures of ferric carboxymycobactin bound by Scn have provided a rationale for the binding trend. Ferric carboxymycobactin structures for n = 5, 6, 7 were crystallized within the Scn calyx. Common to all structures, the hydroxyphenyl-oxazoline of all carboxymycobactins resides in pocket 1 and serves as the main site of Scn affinity due to it being the sole aromatic moiety in carboxymycobactins. The protein structures co-crystallized with ferric carboxymycobactins also revealed that the fatty acid chain length of carboxymycobactins adopts either a 'tail-in' or 'tail-out' (Fig. 4.3) conformation within pocket 2. This conformation change has the most drastic effect on Scn affinity. The carboxylate tail, when in 'tail-in' conformation, creates a crucial interaction between calyx and ligand and thus helps carboxymycobactins reside in the Scn calyx with higher affinity. By this same token, carboxymycobactin fatty acid tails which are too short (n = 3) cannot maintain this interaction with the binding pocket, thus resulting in less favorable binding to Scn. The structure of Scn: carboxymycobactin (n = 5) has previously been solved with the carboxymycobactin tail being both tail-in and tail-out in different crystal structures (Fig. 4.3) [10].

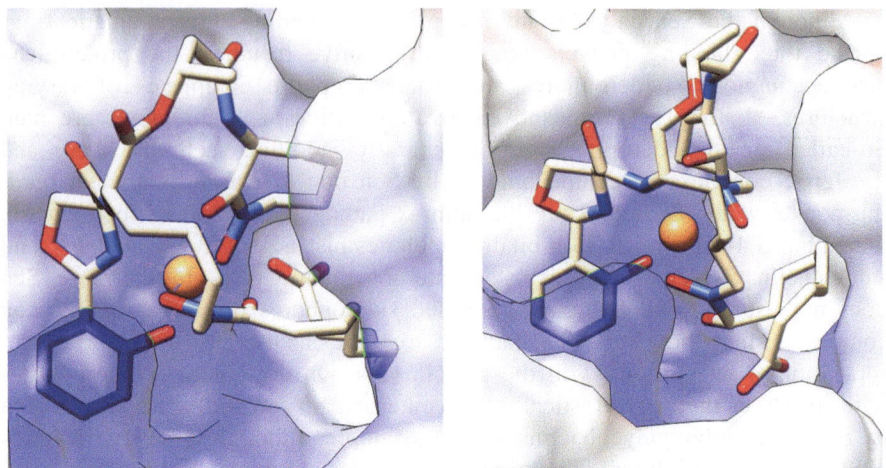

Fig. 4.3 Siderocalin-carboxymycobactin structures. Scn-carboxymycobactin (n = 5) *tail-in* configuration (*left*) and *tail-out* configuration (*right*). Iron is represented by the *orange sphere*; for clarity iron bonds to chelating moieties are not shown

4.4 Scn Location and Contact with Mycobacteria

In many cases, the binding interactions between carboxymycobactin and Scn serve to protect the host from mycobacterial infections in vivo. The observed defense is always iron-dependent, [43, 44] implying that Scn functions through the binding interactions quantified in vitro to intercept ferric carboxymycobactin and inhibit mycobacterial iron acquisition. However, the iron-withholding defense is penetrable, and at least in some cases, mycobacteria have the upper hand in the struggle for iron.

Scn provides significant defense against *M. tuberculosis* at the early infection stage when the pathogen is extracellular. Depending on the location of the infection, different cell types provide protection against an infection by secreting Scn. Alveolar epithelial cells secrete Scn into the alveolar space in response to early mycobacterial infection [44]. Neutrophils secrete Scn and protect against *M. tuberculosis* infections in whole blood alluding to at least one reason why successful *M. tuberculosis* infections in humans have been correlated to a lower neutrophil count [43]. In mice, the Scn level in blood is elevated at least 10 fold in the first two days after infection with *Mycobacterium avium* resulting in a 2–10 fold decrease of bacteria in the blood compared to Scn-knockout mice [35]. When Scn is able to interact directly with mycobacteria in the extracellular environment, it effectively inhibits growth by limiting iron for the pathogen.

After the early stages of a mycobacterial infection, the pathogen moves inside host cells by phagocytosis. Alveolar epithelial cells are infected by mycobacteria, but the number of mycobacteria in alveolar epithelial cells is generally low due to the effectiveness of Scn protection. Significantly more mycobacteria inhabit epithelial cells without Scn than inhabit wild-type epithelial cells expressing Scn in response to the infection. The protection comes as the secreted Scn is internalized and then co-localizes with the mycobacteria to inhibit growth [44]. Scn activity prevents mycobacteria from infecting epithelial cells.

In addition to epithelial cells, alveolar macrophages also produce Scn in response to mycobacterial infection [6, 35, 44]. The effect of Scn in macrophages remains unclear and is highly dependent on the species of mycobacteria. Murine macrophages infected with *Mycobacterium bovis*-bacillus Calmette-Guerin (BCG) permitted similar bacterial growth with or without Scn production [6, 44], but the intracellular *M. tuberculosis* was significantly decreased [6]. *M. tuberculosis* infects an equal number of macrophages whether or not Scn is present, but when Scn is present, fewer bacteria inhabit each macrophage. The bacteria in the wild-type macrophage are elongated and filamentous, indicating that they are growth arrested [6].

To protect against an intracellular pathogen, Scn must enter the macrophage to interact with the mycobacteria. Wild-type Scn is always secreted, but it reenters the cell by endocytosis. In the case of *M. tuberculosis* protection mentioned above, a cytosolic mutant of human Scn lacking the secretion sequence and the wild type protein equally protect a macrophage cell line against *M. tuberculosis* infection [6]. However, *M. avium* uses the intracellular trafficking system of bone marrow-derived macrophages to avoid Scn. Once inside the cell, the *M. avium* places the phagosome in the endocytic recycling compartments which intersect with the ferric transferrin recycling pathway. Endocytosed Scn progress to the lysosomal pathway and cannot interact with the *M. avium* [35]. This species of *Mycobacterium* effectively use the intracellular environment to overcome the host defense and accumulate iron for growth and further infection. Additional work will be helpful to better define the activity of Scn in macrophages during a mycobacterial infection and justify the differences observed between *M. tuberculosis*, *M. avium*, and BCG.

4.5 Concluding Remarks

The battle for iron between humans and pathogens is ancient in development and current in impact on human health. Humans have developed Scn to defend against siderophore-mediated iron acquisition, and some pathogens have responded by making stealth siderophores to avoid Scn recognition. The response by pathogenic mycobacteria appears to be at an intermediate stage. Hydroxamate iron-binding units of carboxymycobactin and mycobactin limit the affinity of Scn, but the aromatic hydroxyphenyl-oxazoline subunit provides an effective handle for Scn recognition. Some alkyl chain lengths prevent Scn binding, while others allow it. Depending on the species, mycobacteria evade Scn protection in alveolar macrophages but are vulnerable in epithelial cells. The intracellular trafficking and location of mycobacteria influences the susceptibility to Scn defense. Scn is part of the human defense that prevents or limits most human infections of *M. tuberculosis*, but limits of this defense are underscored by the large amount of sickness and death that continue to be caused by *M. tuberculosis*.

References

1. Strong RK (2006) Siderocalins. In: Akerstrom B, Borregaard N, Flower D, Salier J-P (eds) Lipocalins. Landes Bioscience, Georgetown
2. Chakraborty S, Kaur S, Guha S et al (2012) The multifaceted roles of neutrophil gelatinase associated lipocalin (NGAL) in inflammation and cancer. Biochim Biophys Acta 1826:129–169
3. Schmidt-Ott KM, Mori K, Li JY et al (2007) Dual action of neutrophil gelatinase-associated lipocalin. J Am Soc Nephrol 18:407–413
4. Goetz DH, Holmes MA, Borregaard N et al (2002) The neutrophil lipocalin NGAL is a bacteriostatic agent that interferes with siderophore-mediated iron acquisition. Mol Cell 10:1033–1043
5. Flo TH, Smith KD, Sato S et al (2004) Lipocalin 2 mediates an innate immune response to bacterial infection by sequestrating iron. Nature 432:917–921
6. Johnson EE, Srikanth CV, Sandgren A et al (2010) Siderocalin inhibits the intracellular replication of *Mycobacterium tuberculosis* in macrophages. FEMS Immunol Med Microbiol 58:138–145
7. Berger T, Togawa A, Duncan GS et al (2006) Lipocalin 2-deficient mice exhibit increased sensitivity to *Escherichia coli* infection but not to ischemia-reperfusion injury. Proc Natl Acad Sci USA 103:1834–1839
8. Abergel RJ, Wilson MK, Arceneaux JEL et al (2006) Anthrax pathogen evades the mammalian immune system through stealth siderophore production. Proc Natl Acad of Sci USA 103:18499–18503
9. Abergel RJ, Zawadzka AM, Raymond KN (2008) Petrobactin-mediated iron transport in pathogenic bacteria: coordination chemistry of an unusual 3,4-catecholate/citrate siderophore. J Am Chem Soc 130:2124–2125
10. Hoette TM, Clifton MC, Zawadzka AM et al (2011) Immune interference in *Mycobacterium tuberculosis*: intracellular iron acquisition through siderocalin recognition of carboxymycobactins. ACS Chem Biol 6:1327–1331
11. Bao G, Clifton M, Hoette TM et al (2010) Iron traffics in circulation bound to a siderocalin (Ngal)–catechol complex. Natlure Chem Biol 6:602–609
12. Correnti C, Strong RK (2012) Mammalian siderophores, siderophore-binding lipocalins, and the labile iron pool. J Biol Chem 287:13524–13531
13. Correnti C, Richardson V, Sia AK et al (2012) Siderocalin/Lcn2/NGAL/24p3 does not drive apoptosis through gentisic acid mediated iron withdrawal in hematopoietic cell lines. PLoS ONE 7:e43696. doi:10.1371/journal.pone.0043696
14. Hoette TM, Abergel RJ, Xu J et al (2008) The role of electrostatics in siderophore recognition by the immunoprotein siderocalin. J Am Chem Soc 130:17584–17592
15. Yang J, Goetz D, Li J-Y et al (2002) An iron delivery pathway mediated by a lipocalin. Molec Cell 10:1045–1056
16. Alpízar-Alpízar W, Laerum OD, Illemann M et al (2009) Neutrophil gelatinase-associated lipocalin (NGAL/Lcn2) is upregulated in gastric mucosa infected with *Helicobacter pylori*. Virchows Arch 455:225–233
17. Ziegler S, Röhrs S, Tickenbrock L et al (2007) Lipocalin 24p3 is regulated by the Wnt pathway independent of regulation by iron. Cancer Gen Cytogen 174:16–23

18. Sunil VR, Patel KJ, Nilsen-Hamilton M et al (2007) Acute endotoxemia is associated with upregulation of lipocalin 24p3/Lcn2 in lung and liver. Exp Molec Path 83:177–187
19. Kjeldsen L, Johnsen AH, Sengeløv H et al (1993) Isolation and primary structure of NGAL, a novel protein associated with human neutrophil gelatinase. J Biol Chem 268:10425–10432
20. Manabe YC, Saviola BJ, Sun L, Murphy JR, Bishai WR (1999) Attenuation of virulence in *Mycobacterium tuberculosis* expressing a constitutively active iron repressor. Proc Natl Acad Sci USA 96:12844–12848
21. Ratledge C (2004) Iron, mycobacteria and tuberculosis. Tuberculosis 84:110–130
22. Gobin J, Moore CH, Reeve JR Jr et al (1995) Iron acquisition by *Mycobacterium tuberculosis*: isolation and characterization of a family of iron-binding exochelins. Proc Natl Acad Sci USA 92:5189–5193
23. Hu J, Miller MJ (1997) Total synthesis of a Mycobactin S, a siderophore and growth promoter of *Mycobacterium smegmatis*, and determination of its growth inhibitory activity against *Mycobacterium tuberculosis*. J Am Chem Soc 119:3462–3468
24. Snow GA (1954) Mycobactin. A growth factor for *Mycobacterium johnei*. III Degradation and tentative structure. J Chem Soc 4080–4093
25. Lane SJ, Marshall PS, Upton RJ et al (1995) Novel extracellular mycobactins, the carboxymycobactins from *Mycobacterium avium*. Tetrahedron Lett 36:4129–4132
26. Krithika R, Marathe U, Saxena P et al (2006) A genetic locus required for iron acquisition in *Mycobacterium tuberculosis*. Proc Natl Acad Sci USA 103:2069–2074
27. Quadri LEN, Sello J, Keating TA et al (1998) Identification of a *Mycobacterium tuberculosis* gene cluster encoding the biosynthetic enzymes for assembly of the virulence-conferring siderophore mycobactin. Chem Biol 5:631–645
28. McMahon MD, Rush JS, Thomas MG (2012) Analyses of MbtB, MbtE, and MbtF suggest revisions to the mycobactin biosynthesis pathway in *Mycobacterium tuberculosis*. J Bacteriol 194:2809–2818
29. Gold B, Rodriguez GM, Marras SAE et al (2001) The *Mycobacterium tuberculosis* IdeR is a dual functional regulator that controls transcription of genes involved in iron acquisition, iron storage and survival in macrophages. Mol Microbiol 42:851–865
30. De Voss JJ, Rutter K, Schroeder BG et al (2000) The salicylate-derived mycobactin siderophores of *Mycobacterium tuberculosis* are essential for growth in macrophages. Proc Natl Acad Sci USA 97:1252–1257
31. Gobin J, Horwitz MA (1996) Exochelins of *Mycobacterium tuberculosis* remove iron from human iron-binding proteins and donate iron to mycobactins in the M. tuberculosis cell wall. J Exp Med 183:1527–1532
32. Olakanmi O, Schlesinger LS, Ahmed A et al (2002) Intraphagosomal *Mycobacterium tuberculosis* acquires iron from both extracellular transferrin and intracellular iron pools. J Biol Chem 277:49727–49734
33. Olakanmi O, Schlesinger LS, Ahmed A et al (2004) The nature of extracellular iron influences iron acquisition by *Mycobacterium tuberculosis* residing within human macrophages. Infect Immun 72:2022–2028
34. Clemens DL, Horwitz MA (1996) The *Mycobacterium tuberculosis* phagosome interacts with early endosomes and is accessible to exogenously administered transferrin. J Exp Med 184:1349–1355
35. Halaas Ø, Steigedal M, Haug M et al (2010) Intracellular *Mycobacterium avium* intersect transferrin in the Rab11$^+$ recycling endocytic pathway and avoid lipocalin 2 trafficking to the lysosomal pathway. J Infect Dis 201:783–792
36. Rodriguez GM, Smith I (2006) Identification of an ABC transporter required for iron acquisition and virulence in *Mycobacterium tuberculosis*. J Bacteriol 188:424–430
37. Stephenson MC, Ratledge C (1980) Specificity of exochelins for iron transport in three species of mycobacteria. J Gen Microbiol 116:521–523
38. Ratledge C, Dover LG (2000) Iron metabolism in pathogenic bacteria. Ann Rev Microbiol 54:881–941

39. Luo M, Fadeev EA, Groves JT (2005) Mycobactin-mediated iron acquisition within mac-rophages. Nature Chem Biol 1:149–153
40. Goetz DH, Willie ST, Armen RS et al (2000) Ligand preference inferred from the structure of neutrophil gelatinase associated lipocalin. Biochemistry 39:1935–1941
41. Gasymov OK, Abduragimov AR, Glasgow BJ (2012) Cation-π interactions in lipocalins: structural and functional implications. Biochemistry 51:2991–3002
42. Scott DC, Newton SMC, Klebba PE (2002) Surface loop motion in FepA. J Bacteriol 184:4906–4911
43. Martineau AR, Newton SM, Wilkinson KA, Kampmann B, Hall BM, Nawroly N, Packe GE, Davidson RN, Griffiths CJ, Wilkinson RJ (2007) Neutrophil-mediated innate immune resist-ance to mycobacteria. J Clin Invest 117:1988–1994
44. Saiga H, Nishimura J, Kuwata H et al (2008) Lipocalin 2-dependent inhibition of mycobacte-rial growth in alveolar epithelium. J Immunol 181:8521–8527

Chapter 5
Siderophore-Mediated Iron Acquisition: Target for the Development of Selective Antibiotics Towards *Mycobacterium tuberculosis*

Raúl E. Juárez-Hernández, Helen Zhu and Marvin J. Miller

Abstract This chapter reviews recent pertinent literature on the *Mycobacterium tuberculosis* siderophore mycobactin and its excreted counterpart carboxymycobactin. Emphasis is placed on the design of antibiotics to specifically interfere with the biosynthesis of these siderophores and the use of siderophore analogs or conjugates to achieve inhibition of *M. tuberculosis*. Although the discussion is focused on biological activity of potential anti-tuberculosis agents, a brief description of the synthetic routes for compounds of interest is given.

Keywords *Mycobacterium* *tuberculosis* • TB • Mycobactins • Siderophore • Siderophore-antibiotic conjugates • Antibiotics • Sideromycins • MDR (multi-drug resistant) • XDR (extensively-drug resistant) • Biosynthesis inhibitors

5.1 Introduction

Approximately one out of every three people living on this planet is infected with *M. tuberculosis,* the causative agent of the contagious disease tuberculosis (TB). It is estimated that in 2009, close to 1.7 million of people died because of this infection [1]. Similar to the emergence of other antibiotic-resistant bacteria, the diagnosis of multi-drug resistant (MDR) and extensively-drug resistant (XDR) strains of *M. tuberculosis* [2] has sparked interest around the world to develop new agents that can be used to eradicate disease caused by this deadly microorganism [3–11].

R. E. Juárez-Hernández (✉) · H. Zhu · M. J. Miller
Department of Chemistry and Biochemistry, University of Notre Dame, Notre Dame, IN, USA
e-mail: rxj159@case.edu

M. J. Miller
e-mail: mmiller1@nd.edu

B. R. Byers (ed.), *Iron Acquisition by the Genus Mycobacterium,*
SpringerBriefs in Biometals, DOI: 10.1007/978-3-319-00303-0_5,
© The Author(s) 2013

Although several approaches can be considered to address the design of novel and selective agents, extensive research has validated the bacterial iron-uptake systems as a biological target for this purpose [12–16, 17]. In this work we describe efforts leading to *M. tuberculosis*-specific (anti-TB) antibiotics by exploiting its iron acquisition system.

5.2 Bacterial Iron Acquisition

Biological systems require iron. Although it is the fourth most abundant element on Earths' crust, its bioavailability is limited by the formation of insoluble salts in aqueous solutions ubiquitous in living organisms. The levels of ferric iron in solution at pH 7 (10^{-9}–10^{-10} M) are considerably lower than the requirements for bacterial growth (10^{-6}–10^{-7} M) [18]. The small and highly charged ferric ion binds strongly to enzymes and functions as a catalyst in a variety of metabolic processes, stressing its importance in life. The human body tightly regulates the iron pools through the action of transferrin in plasma and lactoferrin in secretions. These molecules further reduce the levels of iron (10^{-15}–10^{-25} M) in our bodies [19, 20]. The nutrient-depleted environment limits the ability of invading micro-organisms to establish an infection and in order to satisfy their iron requirements, bacteria and fungi secrete relatively small, high affinity ferric-chelators called siderophores (Fig. 5.1). Hundreds of these molecules have been isolated and characterized [21–25].

While siderophores are structurally diverse, the constituent functional groups involved in iron binding are characteristically conserved: hydroxamic acids, catechols, α-hydroxy acids and aryl oxazolines. While the structure of certain siderophores like desferrioxamine B (1) can be recognized by different bacteria, highly functionalized siderophores like pyoverdin type I (5) can only be acquired by the producing organism. The energetic expense of biosynthesizing these molecules highlights the importance of iron acquisition in bacterial survival. The selective uptake of certain ferri-siderophores by bacteria of interest e.g., pathogenic strains, has sparked extensive research focused on antibiotic delivery through the design of nature-inspired antibiotic conjugates [13, 15, 26], and even bacterial detection through immobilized siderophores [27-28]. The chemical modification and even total syntheses of these derivatives is a challenging task. However, the assembly of remarkably selective and highly potent antibacterial agents validates the efficacy of this field of study.

Siderophore biosynthesis is a highly regulated process (Fig. 5.2). Within an iron-sufficient environment the chelator assembly is repressed at the genetic level by the action of a Fur-Fe^{2+} complex (1). Reduced levels of iron allow the sidero-phore biosynthesis machinery (SBM) to synthesize and secrete the chelator into the extracellular environment (2). The siderophore acquires ferric iron from the environment or, in the case of an infection, the host's pools. After a ferric-complex has been formed, it is recognized by an outer membrane (OM) receptor (3) in the

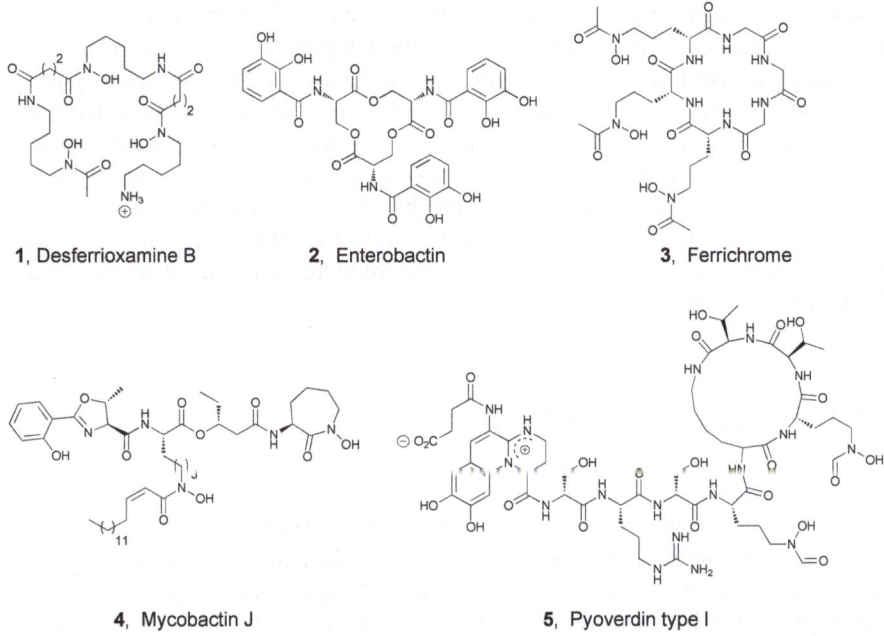

1, Desferrioxamine B   **2**, Enterobactin   **3**, Ferrichrome

4, Mycobactin J   **5**, Pyoverdin type I

Fig. 5.1   Siderophores: structurally diverse, ferric ion-selective chelators [24, 61]

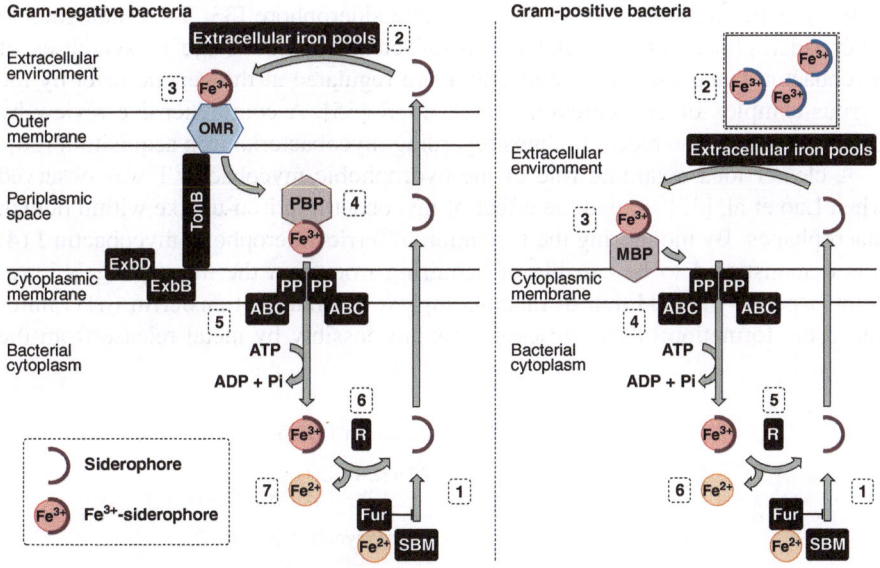

Fig. 5.2   Siderophore-mediated iron uptake in bacteria. Although mycobacterial Fur-like proteins are known, IdeR (a DtxR family member) regulates siderophore expression in the mycobacteria [35]. Adapted from [21, 61]

case of Gram-negative bacteria (left panel). The action of a TonB/ExbB/ExbD system allows the ferri-siderophore to access the periplasmic space where it is recognized by a periplasmic binding protein (PBP) that assists in the delivery to the cytoplasm (4) in an ATP-dependent process (5). Iron release from the siderophore may occur by metal reduction to its ferrous form, which displays a reduced affinity with the chelator. Degradation of the ferric-siderophore can also occur in order to release the metal. Siderophore-acquisition by Gram-positive bacteria occurs is a comparable fashion (right panel), where the ATP-Binding Cassette (ABC) proteins involved in the active transport of ferric-siderophores are related to the system found in Gram-negative bacteria [21–25].

5.3 Siderophore System in *M. tuberculosis*

To acquire iron, *M. tuberculosis* secretes two structurally-related siderophores (Fig. 5.3). Mycobactin T (**6**) discovered by Snow in 1965, is characterized by the presence of a long hydrophobic alkyl chain [29, 30]. Initially proposed to function as a membrane-bound, temporary Fe^{3+}-storage, recent evidence indicates that Mycobactin T is a more active player in the iron-uptake by *M. tuberculosis* [31, 32]. The second chelator, carboxymycobactin T (**7**), was isolated in 1995 by Gobin from virulent *M. tuberculosis* ATCC 35801 and avirulent *M. tuberculosis* $H_{37}Ra$. Initially referred to as exochelin, its structure was found to be similar to that of mycobactin T albeit with a shorter alkyl chain and the presence of an ionizable group, rendering a more water-soluble siderophore [33, 34]. Analogous to other siderophore systems, under iron-sufficient conditions the biosyntheses of mycobactin T and carboxymycobactin T are regulated at the genetic level by the ferrous-complex of the repressor protein IdeR [35]. A comprehensive review by Quadri describes the recent findings regarding mycobacterial iron acquisition [36].

A clearer idea regarding role of the hydrophobic mycobactin T was observed when Luo et al. [32] studied the effect of mycobactin in iron-uptake within human macrophages. By monitoring the formation of ferric-siderophore, mycobactin J (**4**) was demonstrated to be capable of acquiring iron from the macrophage's intracellular pools. The addition of metal-complexes of human transferrin (hTf) influenced the formation of the ferric-siderophore possibly by metal release from the

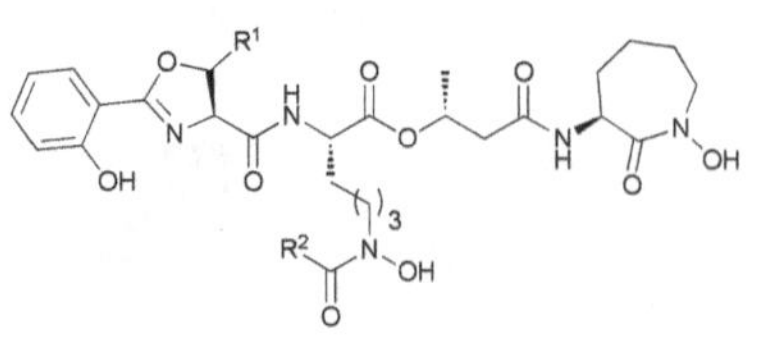

6, Mycobactin T (MbT)
$R^1 = H$
$R^2 = (CH_2)_nCH_3$, n = 16-19;
$(CH_2)_xCH=CH(CH_2)_yCH_3$, x+y = 14-17

7, Carboxymycobactin T (CbT)
$R^1 = H, CH_3$
$R^2 = (CH_2)_nCO_2CH_3/CO_2H$, n = 1-7;
$(CH_2)_xCH=CH(CH_2)_yCO_2CH_3/CO_2H$, x+y = 1-5

Fig. 5.3 Siderophores secreted by *M. tuberculosis* [34, 61, 62]

macrophage's tranferrin receptors, since mycobactin is not efficient in removing iron from transferrin. Both mycobactin J and its ferric-complex were recovered from the cell-aqueous phase. However, fluorescent confocal microscopy showed that although mycobactin J freely diffuses between the phagosome and the intra-cellular environment, Fe^{3+}-MbJ generally is localized in lipid droplets of the cell wall [32]. Work performed by Gobin showed that carboxymycobactin T was capable of removing iron from 40 % saturated hTf, the approximate levels found in human serum. Carboxymycobactin T can also remove the metal from lactoferrin and ferritin although at a slower rate. Potentially operating as a cooperative system, carboxymycobactin T can serve as a source of iron for mycobactin T [33, 34].

Information regarding the role of carboxymycobactin T was obtained through the identification of the genes irtA and irtB. Rodriguez et al. studied the role of the transporter IrtAB in *M. tuberculosis*. Without affecting siderophore biosynthesis, selective mutation interfered with the bacterial ability to use ferric-carboxymyco-bactin T. Under iron-sufficient conditions, the irtAB mutants did not display growth deficiencies. While single mutations (irtA, irtB) were better tolerated, the double irtAB mutant was incapable of growing in iron-deficient conditions. Remarkably, ferric-carboxymycobactin T was used efficiently without the assistance of myco-bactin T by a mycobactin mutant of *M. tuberculosis*, a strain incapable of synthesiz-ing siderophores [35, 37]. It is not clear if mycobactin T uses the IrtAB transporter or if it diffuses through the cell envelope due to its hydrophobicity. In any case, similar to other bacterial siderophore systems, the release of iron from the myco-bactin T or carboxymycobactin T in *M. tuberculosis*, could take place by metal reduction to its ferrous form. Evidence for this was obtained when Ratledge [38] studied the reductase-triggered iron release from ferric-mycobactin S in *M. smeg-matis* [38–40]. Details about iron delivery into the mycobacterial cytoplasm and the exact mechanism of acquisition in vivo are still the subject of current research.

5.4 Siderophore-Antibiotic Conjugates

The importance of iron has prompted bacteria to develop the ability to recog-nize exogenous ferric-siderophores in order to obtain a competitive advantage. Sideromycins are naturally occurring siderophore-antibiotic conjugates (Fig. 5.4) synthesized by a variety of bacteria to counter the thievery of metallic complexes. The albomycins (8) are hydroxamate-based sideromycins containing a lethal seryl-t-RNA synthetase inhibitor that is released by the activity of a serine pepti-dase N in some bacteria. The inability of other bacteria to actively transport the ferric-albomycin and the lack of the enzymatic activity to release the antibiotic moiety, results in no inhibitory activity by these compounds [41, 42]. Microcin MccE492m (9) is an enterobactin-based, peptide antibiotic-conjugate. Microcins are pH and heat stable peptides secreted by enterobacteria under nutrient restric-tive environments. MccE492m, secreted by *K. pneumonaie* RYC492, displays potent and specific inhibitory activity (MIC = 0.2–1.2 μM) against *Escherichia*

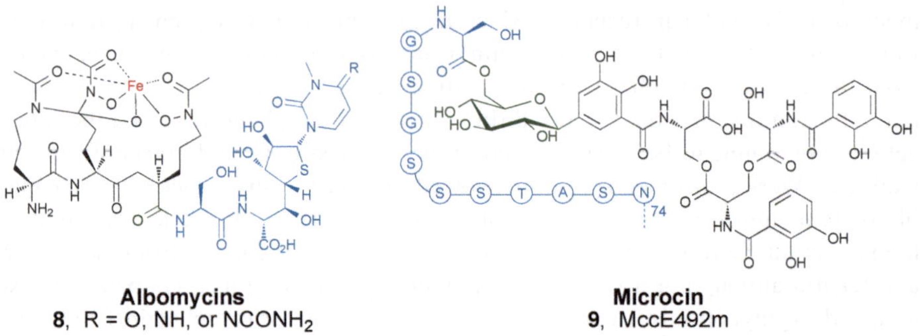

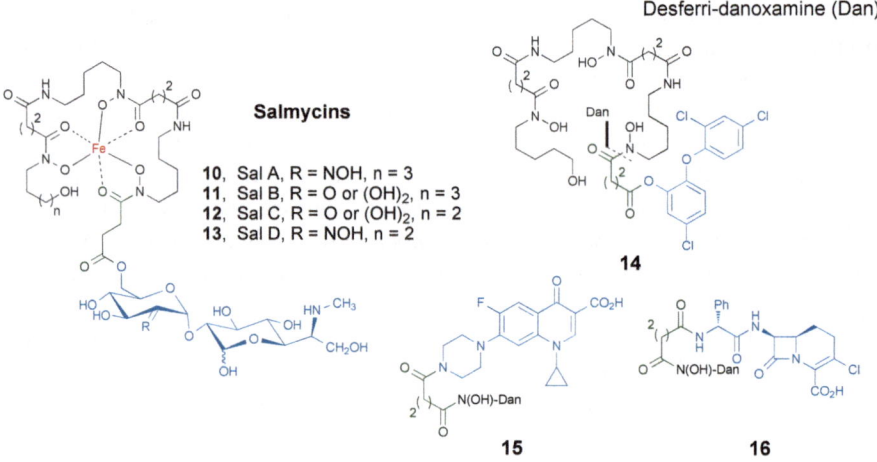

Fig. 5.4 Examples of naturally occurring sideromycins. The antibiotic moiety is highlighted in *blue*, the linker in *green* [61]

Fig. 5.5 Synthetic sideromycins inspired by the natural occurring Salmycins. The antibiotic moiety is highlighted in *blue*, the linker in *green* [48, 51, 61]

and *Salmonella* spp. MccE492 enters the bacteria through the siderophore-uptake system and disrupts the cytoplasmic membrane while interacting with other intracellular proteins [43–46].

Salmycin B (**11**, Fig. 5.5) represents another example of a naturally occurring sideromycin. Isolated by Vértesy from *Streptomyces violaceous* DSM 8286, this and other related salmycins (**10–13**) are potent growth inhibitors of Staphylococci and Streptococci [47]. The assembly of the hydroxamate-based siderophore portion, danoxamine (Dan), allowed the complete synthesis of desferrisalmycin B [48, 49]. The availability of synthetic fragments aided the correct stereochemical assignment of the antibiotic moiety. Wencewicz explored the use of Dan to deliver different antibiotics (**14–16**) in an attempt to mimic the nature-inspired activity of sideromycins [50, 51]. The biological activity of synthetic sideophore-conjugates provided evidence that selective and, in certain cases, potent inhibitors can be developed by target the iron-uptake as a target [52].

5.5 Mycobactin Analogs are Growth Inhibitors of *M. tuberculosis*

G. A. Snow performed extensive siderophore work with the isolation and characterization of the mycobactins. The early realization that exogenous mycobactins could disrupt mycobacterial growth led to the study and chemical syntheses of analogs in search of potential inhibitors of *M. tuberculosis* (Fig. 5.6) [30]. Maurer and Miller reported the synthesis of mycobactin S2 (**17**, MbS2), the first example of an iron binding capable synthetic mycobacterial siderophore [53]. However, **17** did not display growth inhibitory activity against *M. tuberculosis*. The lack of a long alkyl chain proved to be important for biological activity, as demonstrated with the synthesis of MbS (**18**), which effectively inhibited *M. tuberculosis* H37Rv (MIC99 = 12.5 µg/mL, 15.6 µM). Stereochemistry at the ester region of the mycobactin core was proven to be critical. While **18** was a potent inhibitor of *M. tuberculosis*, the synthetic (*R*)-epimer mycobactin T (**19**) was a growth promoter [54]. Structurally related to the mycobactins, amamistatin B is a natural product secreted by the actinomycete *Nocardia asteroides*. Fennell and collaborators reported the syntheses and biological activity of amamistatin B analogs (**20–22**), where only analog **20** was found to display modest growth inhibition against *M. tuberculosis* (MIC = 47 µM) [55].

Pursuing the synthesis of a mycobactin-antibiotic conjugate, Miller et al. [56, 57] reported analog **23** (Fig. 5.7). In order to incorporate a chemical handle for further manipulation, the ester region of the siderophore was modified to contain a 2,3-diaminopropionate spacer. Remarkably, **23** itself, with a *N*-Boc protecting group was found to be a potent inhibitor of *M. tuberculosis* H37Rv (MIC < 0.2 µg/mL)

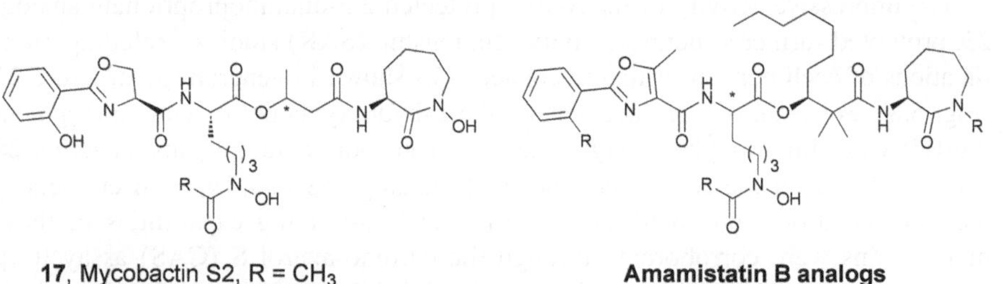

17, Mycobactin S2, R = CH3
18, Mycobactin S, (*S*), R = (CH2)14CH3
19, Mycobactin T, (*R*), R = (CH2)14CH3

Amamistatin B analogs
20, (*R*), R = OH
21, (*S*), R = OH
22, (*S*), R = H

Fig. 5.6 Synthetic mycobactin and amamistatin analogs [61, 55]

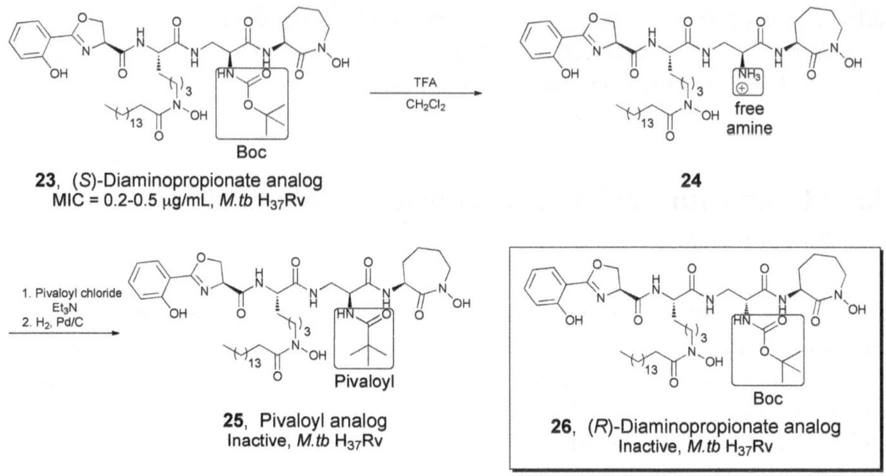

Fig. 5.7   Diaminopropionate modified mycobactin analogs [56, 57, 58]

[56, 57]. However, the free amine **24** obtained by removal of the protecting group for subsequent derivatization and preparation of antibiotic conjugates was found to be completely inactive. Anticipating that most likely the loss of activity was due to the removal of the bulky hydrophobic *N*-Boc group and generation of a positively charged amine that might not be taken up or localized in the same lipophilic region of the mycobacterial cellular envelope, the free amine was reacylated with a pivaloyl group to give derivative **25**. Interestingly, while pivalate derivative **25** differs only by one oxygen relative to the active *N*-Boc protected compound, the pivaloyl analog **25** was devoid of antibiotic activity. No reason for this remarkable difference has been determined but it is interesting to speculate that the active *N*-Boc protected compound might actually serve as a prodrug that is hydrophobically assimilated and acid sensitive Boc group "deprotected" to generate the positively charged amine in the local environment. The pivaloyl derivative **25** would be inert towards such mild deacylation conditions [57, 58].

The impressive activity of the *N*-Boc-protected 2,3-diaminoproprinate analog, **23**, prompted further structure-activity-relationship (SAR) studies, including modifications of each portion of the mycobactins as shown in generalized structure **27** (Fig. 5.8). For example, replacement of the *o*-hydroxyphenyl oxazoline (segment "A–B") with dihydroxybenzoylglycine gave catechol-containing mycobactins **28** and **29**. Based on the results obtained with analogs **18** and **19**, both epimers at the ester position were synthesized [59]. The iron-binding capabilities of these mycobactins were corroborated through the chrome azurol S (CAS) assay [60]. However, both analogs did not display growth inhibition against *M. tuberculosis* (MIC > 6.25 μg/mL, >7.75 μM), perhaps stressing the importance of the oxazoline moiety in biological recognition. Interestingly, **28** and **29** were efficiently used as growth promoters by strains of *M. smegmatis* [59].

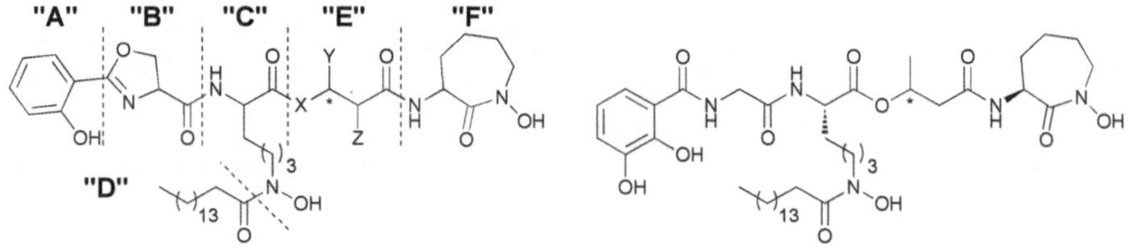

27

Catechol-containing mycobactins
28, (*R*)
29, (*S*)

Fig. 5.8 Generalized mycobactin structructure (**27**), catechol-containing analogs [59]

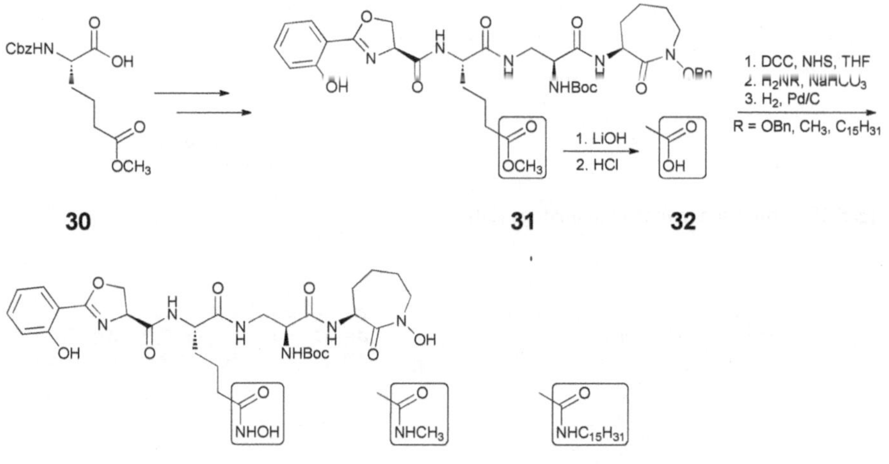

33 **34** **35**

Fig. 5.9 Modification on the component "C" of the mycobactin structure [58]

Changes of the linear lysine hydroxamate component "C" also were not tolerated. Thus, shortening the side chain of **23** to an acetyl as in mycobactin S2 resulted in loss of inhibitory activity against *M. tuberculosis*. Replacement of the linear ε-*N*-hydroxy lysine with an α-amino adipate allowed syntheses of a series of analogs **33–35** (Fig. 5.9) with variation of lipophilicity and iron binding capabilities. Not surprisingly, none of these compounds displayed anti-TB activity.

Since all of the components A–F and related fragment-analogs of the generalized mycobactin structure **27** were synthetically available, a number of "truncated" mycobactins were also synthesized (Fig. 5.10), tested and found to be inactive against *M. tuberculosis* [58].

In the development of a platform for the convergent manipulation of mycobactin templates that retained the core structure of the natural mycobactins, Juárez-Hernández et al., reported the syntheses of mycobactin analogs **47–49** (Fig. 5.11)

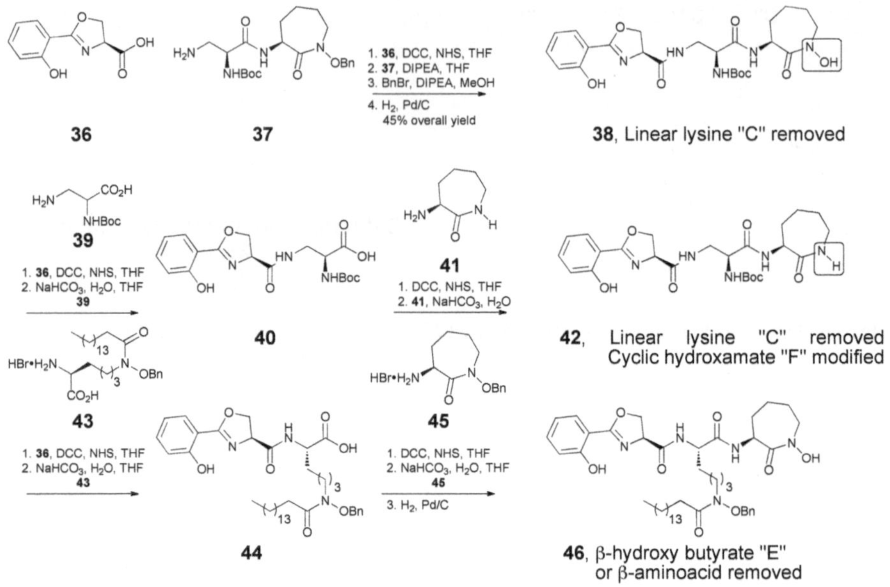

Fig. 5.10 Truncated mycobactin analogs [58]

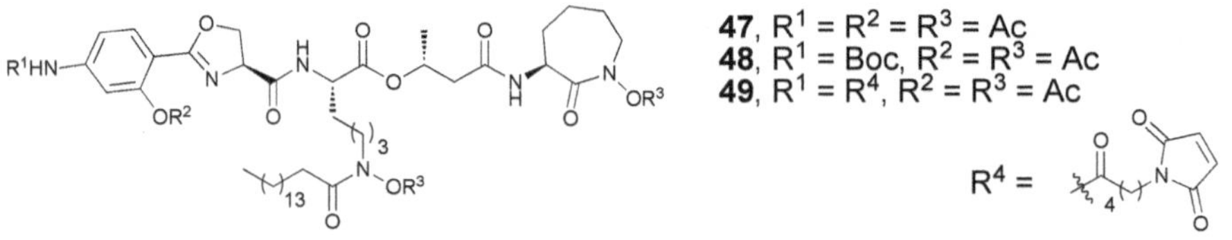

Fig. 5.11 Maleimide-containing mycobactin T analogs [61, 62]

[61, 62]. In order to perform functionalization through the aryl-oxazoline moiety, an amino group and a maleimide linker were incorporated at the phenyl ring. The mycobactin core was not further modified. Screened against replicating *M. tuberculosis*, the synthetic analogs were found to be potent growth inhibitors in the Microplate Alamar Blue Assay (MABA), **47** (MIC = 0.09 μM in 7H12 media, MIC = 0.43 μM in GAS), **48** (MIC = 0.02 μM in 7H12 media, MIC = 2.88 μM in GAS), **49** (MIC = 0.88 μM in 7H12 media, MIC = 1.02 μM in GAS). The analogs were tested in the Low-Oxygen-Recovery-Assay (LORA) to assess activity against non-replicating *M. tuberculosis* (MICs > 50 μM). More studies are necessary to determine if the difference in observed activity between both assays is due to the metabolic needs of the mycobacteria. However, these analogs were found to be specific inhibitors of *M. tuberculosis* as no effect was observed when tested in an agar-diffusion assay against a panel of Gram-positive and Gram-negative bacteria. While the different levels of activity observed in these analogs,

appears to indicate biological discrimination by *M. tuberculosis*, the specifics regarding the level at which recognition occurs is not yet determined. It is important to indicate that the mycobactin analogs must be interfering with the iron acquisition system considering that non metal-binding precursors (*O*-benzyl protected hydroxamates) do not display any antibiotic activity. Although it has been demonstrated that mycobactin T can acquire iron from the lipid droplets inside macrophages and act cooperatively with carboxymycobactin T to recover metal from other sources such as transferrin and lactoferrin, much remains to be learned from these fascinating molecules.

The lack of anti-TB activity of all of the analogs described above, except for the Boc-protected 2,3-diaminoproprionate **23** and the *p*-aminosalicylate (PAS) analogs **47–49**, reflected the growingly apparent remarkable iron transport selectivity of mycobacteria. However, further studies revealed two exciting leads for the development of anti-TB agents based on studies of mycobactins: the discovery of new classes of non-iron metabolism based small heterocycle inhibitors and the ability to utilize **23** for the preparation of mycobactin-like drug conjugates (sideromycins) with exquisitely selective anti-TB activity [56, 57].

5.6 Small Molecule Anti-TB Agents Derived from Studies of the Oxazoline Component of Mycobactins

During the process of screening the synthetic mycobactins, analogs and components, a double benzyl protected form **51** of the *o*-hydroxyphenyl oxazoline component **50** was found to have moderate but very selective inhibitory activity against *M. tuberculosis*, presumably through an iron trafficking independent mechanism since its iron binding constituent was masked. Phenyl oxazoline, **52** without the benzyloxy group was synthesized and tested to confirm that related non-iron binding compounds possessed anti-TB activity. In fact, the activity of **52** was enhanced relative to the dibenzyl protected lead compound (**51**) and, again, the compound was found to be extraordinarily potent against *M. tuberculosis*, while not active against other forms of mycobacteria or other Gram-positive or Gram-negative bacteria. These early results prompted extensive structure-activity relationship (SAR) studies of related hydrophobic oxazoline, oxazole and similar heterocycles. The results are summarized in the generalized structure **53** below (Fig. 5.12) [63, 64-65].

While developing the SAR of the oxazolines, oxazoles and related compounds, many other heterocyclic cores were prepared, derivatized and screened, including imidazopyridines, which have emerged as tremendously potent and selective anti-TB agents [66]. Imidazopyridines have nanomolar anti-TB activity, including similarly potent activity against multi-drug resistant (MDR) and extensive drug resistant (XDR) clinical isolates of *M. tuberculosis*, but are not broadly active against all types of mycobacteria or other Gram-positive or Gram-negative bacteria. The compounds are easily synthesized, appear to be very drug like with ideal

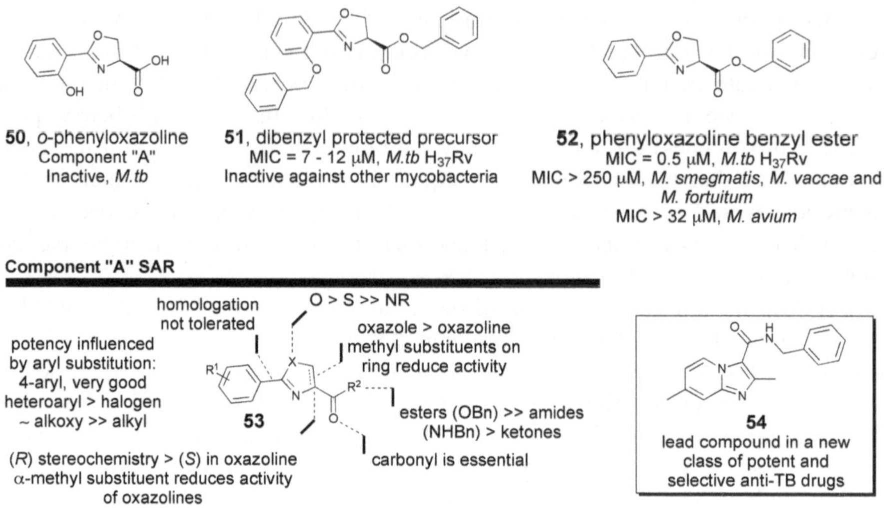

50, o-phenyloxazoline
Component "A"
Inactive, *M.tb*

51, dibenzyl protected precursor
MIC = 7 - 12 μM, *M.tb* H$_{37}$Rv
Inactive against other mycobacteria

52, phenyloxazoline benzyl ester
MIC = 0.5 μM, *M.tb* H$_{37}$Rv
MIC > 250 μM, *M. smegmatis*, *M. vaccae* and
M. fortuitum
MIC > 32 μM, *M. avium*

Component "A" SAR

homologation
not tolerated

O > S >> NR

potency influenced
by aryl substitution:
4-aryl, very good
heteroaryl > halogen
~ alkoxy >> alkyl

oxazole > oxazoline
methyl substituents on
ring reduce activity

53

esters (OBn) >> amides
(NHBn) > ketones

(*R*) stereochemistry > (*S*) in oxazoline
α-methyl substituent reduces activity
of oxazolines

carbonyl is essential

54
lead compound in a new
class of potent and
selective anti-TB drugs

Fig. 5.12 Structure-activity relationship of mycobactin component "A" [63, 64, 65]

molecular weight, Log P, good in vitro stability, are orally bioavailable and have an acceptable therapeutic index with lack of toxicity against human cell lines [63].

5.7 A Mycobactin Siderophore-Antibiotic Conjugate Selectively Targets *M. tuberculosis*

Siderophore-antibiotic drug conjugates have demonstrated the potential to be strong and selective inhibitors of bacteria. Siderophores that are recognized by a unique strain are of particular interest. To explore if mycobactin T analogs could be used to deliver antibiotics into *M. tuberculosis*, Miller et al. reported the assembly of mycobactin-artemisinin conjugate **56** (Fig. 5.13) [56]. Artemisinin is an endo-peroxide containing antibiotic used to treat *P. falciparum*, the causative agent of malaria. The conjugate was proven to be a potent inhibitor of *M. tuberculosis* H$_{37}$Rv (MIC = 0.39 μg/mL), with remarkable inhibition of MDR *M. tuberculosis* (MIC = 0.16–1.25 μg/mL), and XDR *M. tuberculosis* (MIC = 0.078–0.625 μg/ mL). Mycobactin conjugate **56** was found inactive against a panel of Gram-positive and Gram-negative bacteria but inhibited four strains of *P. falciparum* (IC$_{50}$ = 0.004–0.0051 μg/mL). The results were consistent with the fact that the synthetic mycobactin T analogs and the artemisinin core do not display broad antibiotic activity. However, the artemisinin peroxide moiety is sufficient to inhibit *P. falciparum*. To confirm the observed activity, desferrioxamine-artemisinin conjugate **58** was synthesized. As expected, **58** inhibited *P. falciparum* (IC$_{50}$ > 2.2–0.16 μM) but was found to be inactive against *M. tuberculosis* H$_{37}$Rv. This work confirmed that selectivity of sideromycins can be modulated through

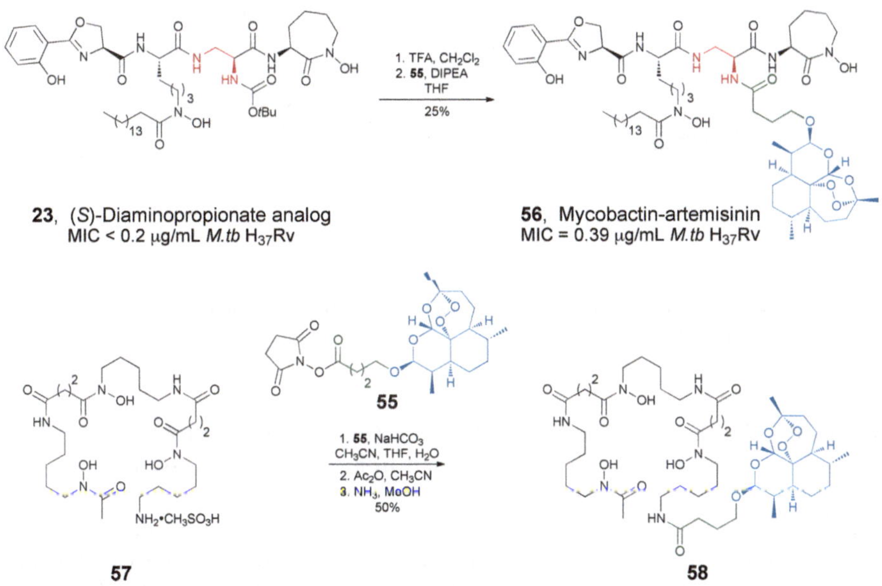

23, (S)-Diaminopropionate analog
MIC < 0.2 μg/mL *M.tb* H$_{37}$Rv

56, Mycobactin-artemisinin
MIC = 0.39 μg/mL *M.tb* H$_{37}$Rv

Fig. 5.13 Mycobactin-artemisinin conjugate. The antibiotic moiety (artemisinin) is highlighted in *blue* [61, 56]

the siderophore portion as demonstrated by the bacterial specificity, or through the antibiotic moiety as observed by artemisinin's spectrum of activity.

5.8 Inhibition of MbtA Nonribosomal Peptide Synthetase

Because the biosynthesis of mycobactin T is essential for the growth of *M. tuberculosis* under iron-limited conditions [67, 68], the enzymes involved in the assembly of the siderophore have become targets for the development of novel antibiotic agents. The biosynthesis of mycobactin T is performed by a mixed nonribosomal peptide synthetase polyketide synthase (NRPS-PKS) assembly line. The genes encoding for the siderophore machinery are clustered in two regions, the mycobactin *T-1* cluster contains the enzymes involved in the synthesis of the mycobactin core, while mycobactin *T-2* performs the activation and incorporation of the hydrophobic tail [69-70]. NRPS are enzymes that catalyze multistep reactions. Organized in modules, NRPS assist the incorporation of amino acid residues unto bacterial metabolites. The A-domain of these enzymes activates a specific substrate to form an acyl adenylate intermediate (**60**, Fig. 5.14) in an ATP-dependent process [71, 72, 73]. Such activity is related to the aminoacyl tRNA synthetases although there is no sequence, nor structural relationship between them. Finking et al. [74], reported the first study to develop A-domain inhibitors through the synthesis of analogs **61** and **62**. These compounds were screened against two

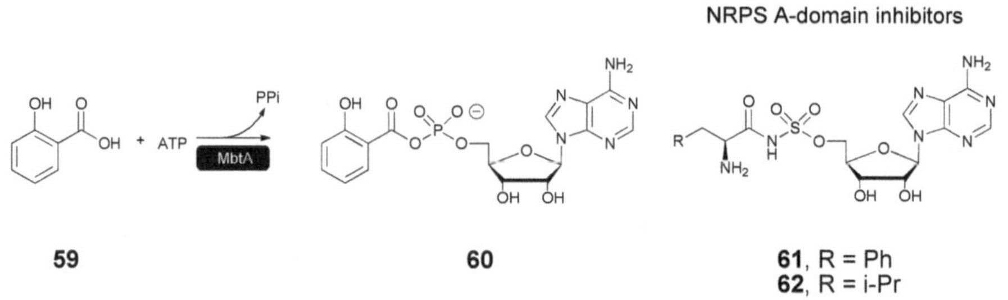

NRPS A-domain inhibitors

59 **60** **61**, R = Ph
62, R = i-Pr

Fig. 5.14 Adenylate-forming activity of A-domains in NRPS (*left*) and examples of selective inhibitors (*right*) [73, 74]

A-domains: GrsA, a phenylalanine-activating (PheA) domain from *Bacillus brevis*, and a leucine-activating (LeuA) domain from *Bacillus subtilis*. The report of the crystal structure of GrsA in the presence of adenosine monophosphate and phenylalanine provided an insight in the mode of action of the A-domain. From the inhibition assays, the presence of **61** inhibited the activity of PheA with L-Phe (K_i = 61 nM) and D-Phe (K_i = 63 nM), while **62** inhibited LeuA with L-Leu (K_i = 8.4 nM). Since no cross-inhibition was observed between domains, this work demonstrated the ability to develop selective NRPS A-domain inhibitors where minor structural changes could afford specificity.

Inspired by the previous work, May et al. [75] demonstrated inhibition of DltA, a carrier protein ligase that selectively activates D-Ala during the D-Ala adenylation of lipoteichoic and teichoic acid in *B. subtilis* and other Gram-positive bacteria. Compound **63** (Fig. 5.15) was found to be a potent inhibitor of DltA isolated from *B. subtilis* (K_i = 232 nM). Because structurally related ascamycin (**64**) is found in fermentation of *Stroptomyces*, it was expected that **63** would display the ability to diffuse through cell membranes. Growth inhibition using *B. subtilis* and a DltA-deletion mutant showed that the phenotype observed in the presence of **63** was consistent with the DltA-deletion mutant. Additional evidence of selectivity was obtained when compound **63** did not inhibit A-domains PheA and DhbE at a concentration of 2 mM.

Ferreras et al. [76] studied the inhibition of adenylate-forming enzymes MbtA from *M. tuberculosis* and YbtE from *Yersinia pestis*. Because other mechanistically-related enzymes had been shown to bind their adenylate intermediates 2–3 times more strongly than the corresponding substrates, it was proposed that compound **65** (Fig. 5.16) could inhibit both MbtA and YbtE. The report of a crystal structure of DhbE with **66** led to the conclusion that **66** would bind in a similar fashion. Recognition appeared to be based primarily on H-bonding and not electrostatic interactions and therefore **65** was expected to be an adequate surrogate. Under iron-limited conditions, compound **65** was demonstrated to inhibit the production of mycobactin in *M. tuberculosis* (IC_{50} = 2.2 ± 0.3 µM) and yersiniabactin in *Y. pestis* (IC_{50} = 51.2 ± 4.7 µM). No inhibition of *Y. pestis* was observed in iron-sufficient medium, while *M. tuberculosis* displayed a 18-fold reduced potency

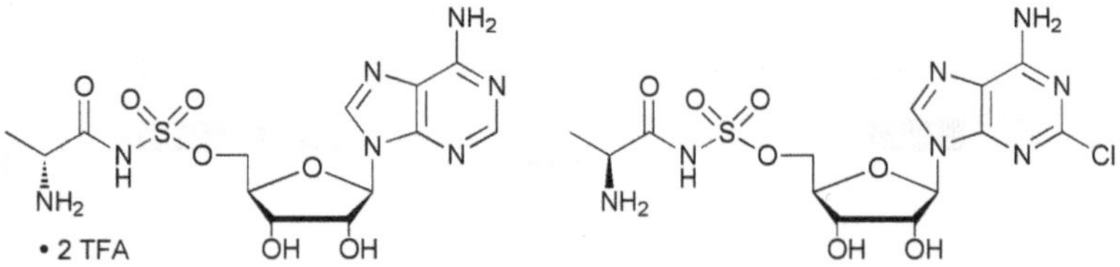

63   **64**, ascamycin

Fig. 5.15 DltA-specific inhibitor (**63**) and ascamycin (**64**) [75]

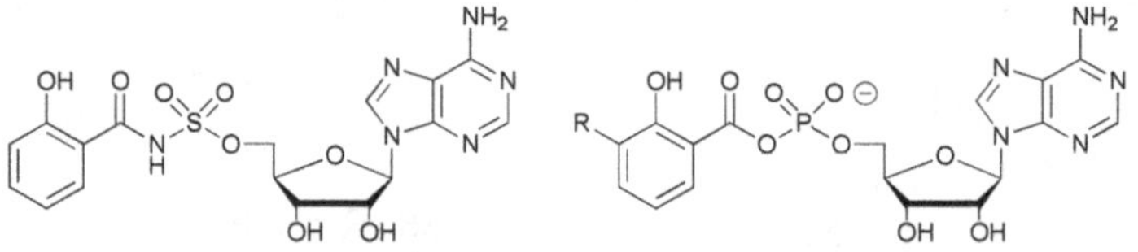

65   **60**, R = H
                                                 66, R = OH

Fig. 5.16 Salicyl-adenylate inhibitor of mycobactin and yersiniabactin [76]

(IC_{50} = 39.9 ± 7.6 μM) perhaps by other mode of action. Compound **65** represents the first biochemically confirmed inhibitor of siderophore biosynthesis.

Somu et al. [73] have further explored the syntheses of nucleoside antibiotics that target MbtA, an adenylate-forming enzyme that activates salicylic acid and loads it onto a thiol group of mycobactin TB in a two-step process (Fig. 5.17). Like other A-domain modules, the lack of mammalian homologues, makes MbtA an attractive target for the development of antibacterial agents. The bisubstrate inhibitors were designed based on adenylate intermediate **60** considering that its A-domain binding was 3–5 orders of magnitude greater than that of the salicylic acid. The tight binding prevents the loss of the intermediate to the surroundings or hydrolysis.

Initial work focused on modification of the phosphate linker and substitution on the salicyl aryl moiety **65**, **67–68** (Fig. 5.18). Inhibitor **65** was designed based on the natural product ascamycin (**64**). The hydroxyl group in the salicyl moiety was demonstrated to modulate the acidity of the NH proton of the linker through delocalization. The acylsulfamate linker proved to be unstable and in order to overcome this limitation, all analogs were prepared as the corresponding triethyl-ammonium salts [73].

The nucleoside inhibitors were screened against *M. tuberculosis* H37Rv under iron-limiting conditions to test affinity, stability and permeability across the mycobacterial cell envelope. Acylsulfamide analog **69** (Fig. 5.18), designed to improve overall stability, displayed the highest inhibitory activity (MIC_{99} = 0.19 μM), comparing favorably with isoniazid (MIC_{99} = 0.18 μM).

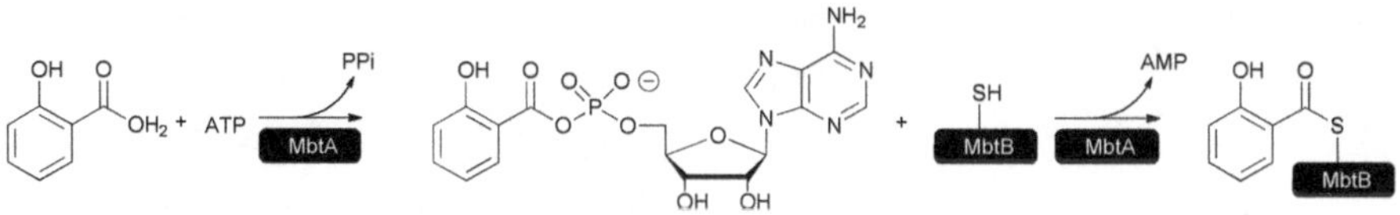

Fig. 5.17 Adenylate-forming reaction catalyzed by MbtA [73]

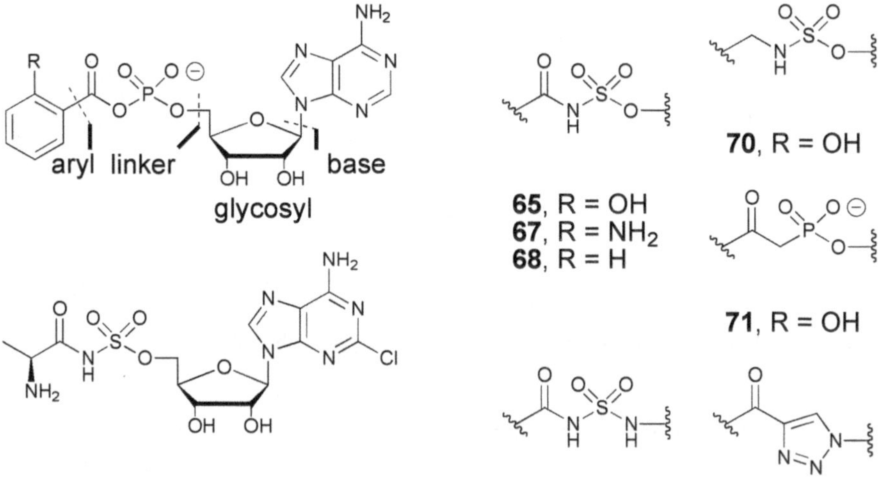

Fig. 5.18 Rationally designed nucleoside antibiotics. SAR of the linker portion [73]

Substitution on the salicyl ring demonstrated the importance of the hydroxyl group. Inhibitor **68** (MIC_{99} = 12.5 μM) displayed a 66-fold decrease in activity over analog **65** (MIC_{99} = 0.29 μM), while the aniline **67** had further diminished activity (MIC_{99} > 100 μM). Inhibitors **70–72** demonstrated no growth inhibition (MIC_{99} > 100 μM). In the case of analog **70**, modeling confirmed the loss of hydrogen bonding with a lysine residue (Lys519). The completely ionized phosphate in analog **71** probably limits permeability through the hydrophobic cell envelope. Compounds **65** and **69** were tested for their ability to block siderophore synthesis. The addition of **65** (20 μM) or **69** (10 μM) completely inhibited the production of both mycobactin T and carboxymycobactin T.

Hydrolysis of the acylsulfamate **65** and acylsulfamide **69** linker present in nucleoside antibiotics releases highly cytotoxic adenosine fragments **75** and **76** (Fig. 5.19). The possibility of this reaction in vivo would limit the application of the inhibitors. To address this concern, Vannada [77] reported the synthesis of **73** and **74** where the central nitrogen was removed. β-Ketosulfonamide analog **73** displayed modest

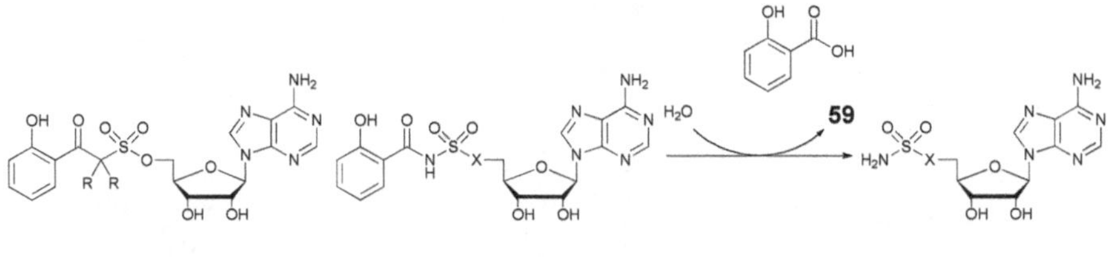

Fig. 5.19 Rationally designed nucleoside antibiotics. Non-hydrolyzable linker [77]

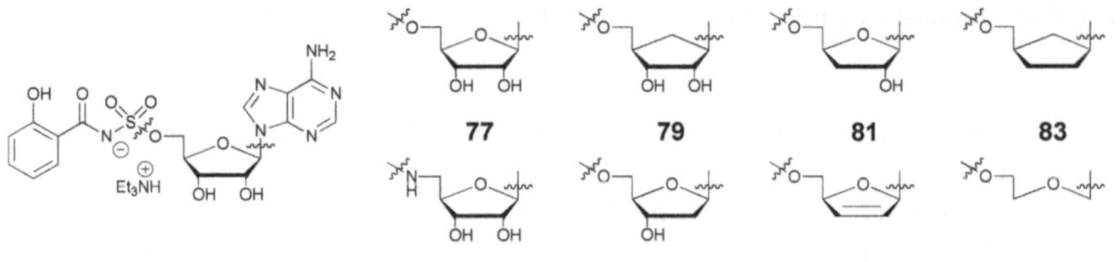

Fig. 5.20 Rationally designed nucleoside antibiotics. SAR of the glycosyl portion [78]

inhibition of MbtA (K_I^{app} = 3.30 + 0.57 μM) and retained moderate anti-TB activity (MIC_{99} = 25 μM), while **74** was found inactive in both assays. Molecular docking with salicyl-adenylating DhbE showed that the analogs deviated from a planar conformation necessary for activity. Compound **75** was also tested and was found to display reduced activity (MIC_{99} = 50 μM), providing evidence that the potent inhibition of **65** and **69** was not due to hydrolysis and release of a cytotoxic byproduct.

In subsequent work, Somu et al. [78] explored variation on the glycosyl region of the nucleoside inhibitors (Fig. 5.20). Carbocyclic analog **79** displayed reduced activity (MIC_{99} = 1.56 μM) compared to the previously synthesized **77** (Et_3N adduct of **65**) and **78** (Et_3N adduct of **69**). The removal of hydroxyl groups in the sugar ring reduced or eliminated all inhibitory activity **80** (MIC_{99} = 25 μM), **81** (MIC_{99} = 1.56 μM), **82** (MIC_{99} > 200 μM). The antibiotics were tested against purified mycobactin TA and in general, the observed inhibitory activity correlated with the whole-cell assays. Among the exceptions, analog **83** was found inactive in vivo (MIC_{99} > 200 μM) while inhibiting MbtA (K_I^{app} = 0.061 μM). This observation was consistent with the idea that these compounds are actively transported that the sugar moiety is important for recognition.

Utilizing molecular modeling, Neres elaborated the SAR of the base moiety of the nucleoside adenylate-inhibitors (Fig. 5.21) [79]. Compounds **102–104** displayed potent growth inhibition against *M. tuberculosis* under iron-deficient (MIC_{99} = 0.049 μM) and iron-sufficient conditions (MIC_{99} = 0.39 μM), representing the most potent compounds of this class reported to date. In general, it was evident that at least one H-bond donor at *N*-6 was essential for activity. Substitution at the *N*-6 position with alkyl groups increased the inhibitory activity

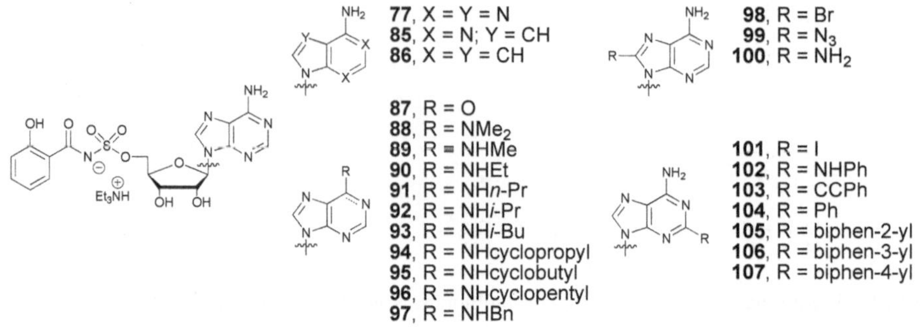

77, X = Y = N
85, X = N; Y = CH
86, X = Y = CH

98, R = Br
99, R = N$_3$
100, R = NH$_2$

87, R = O
88, R = NMe$_2$
89, R = NHMe
90, R = NHEt
91, R = NHn-Pr
92, R = NHi-Pr
93, R = NHi-Bu
94, R = NHcyclopropyl
95, R = NHcyclobutyl
96, R = NHcyclopentyl
97, R = NHBn

101, R = I
102, R = NHPh
103, R = CCPh
104, R = Ph
105, R = biphen-2-yl
106, R = biphen-3-yl
107, R = biphen-4-yl

Fig. 5.21 Rationally designed nucleoside antibiotics. SAR of the base portion [79]

as reflected in the activity of cyclopropyl analog **94**. To analyze the analogs, a selectivity factor S was defined as the ratio of inhibitory activity between iron-sufficient and iron-deficient conditions. Analog **94** demonstrated to be the most promising inhibitor with an S value of 64 (MIC$_{99}$ = 0.098 μM/MIC$_{99}$ = 6.25 μM). The difference between potent MtbA inhibition and whole-cell *M. tuberculosis* may reflect the reduced permeability of these compounds where structural modification affects the recognition and transport of the nucleosides. The analogs did not display cytotoxicity (IC$_{50}$ > 100 μM) and, even more remarkably, parent compound **77** did not inhibit a panel of four FadD adenylating enzymes from *M. tuberculosis*, a demonstration of the exquisite selectivity of these nucleoside antibiotics.

5.9 Inhibition of MbtI Nonribosomal Peptide Synthetase

Work performed by Harrison et al., demonstrated that the gene Rv2386c, encoding for enzyme salicylate synthase MbtI was essential for in vitro growth of *M. tuberculosis*. MbtI catalyzes the production of salicylate **109** and pyruvate **110** from chorismate **108** (Fig. 5.22) in the first committed step during the biosynthesis of mycobactin T [80]. The crystal structure of MbtI and other chorismate-utilizing enzymes, provided yet another target for the development of antibiotic agents targeting *M. tuberculosis*.

Manos-Turvey explored the synthesis of MbtI inhibitors through the synthesis of chorismate analogs **111–113** (Fig. 5.23). The compounds were tested against mycobactin TI and chrosimate-utilizing enzyme *Serratia marcescens* anthranilate synthase (AS) [81]. In general, poor inhibition was observed against MbtI with **111** (K$_i$ = 1400 ± 400 μM), and **112** (K$_i$ = 3000 ± 1000 μM), **113** (K$_i$ = 1700 ± 500 μM), probably due to steric hindrance by the C5 hydroxy substituent as observed in molecular docking. However, the analogs were better inhibitors of *S. marcescens* AS, **111** (K$_i$ = 28 ± 7.0 μM), **112** (K$_i$ = 90 ± 14 μM), and **113** (K$_i$ = 230 ± 40 μM). Simplified analog **114** was synthesized to assess the effect of the hydroxyl substituents and proved to be a better inhibitor against both enzymes MbtI (K$_i$ = 500 ± 90 μM), *S. marcescens* AS (K$_i$ = 3.2 ± 0.3 μM).

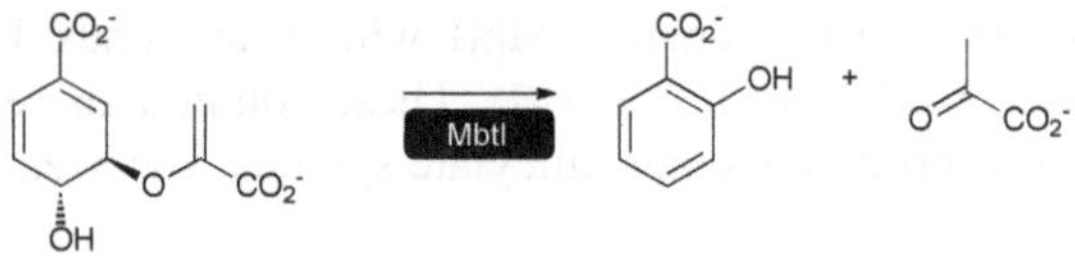

108, chorismate   **109**, salicylate **110**, pyruvate

Fig. 5.22 MbtI-catalyzed synthesis of salicylate from chorismate [80]

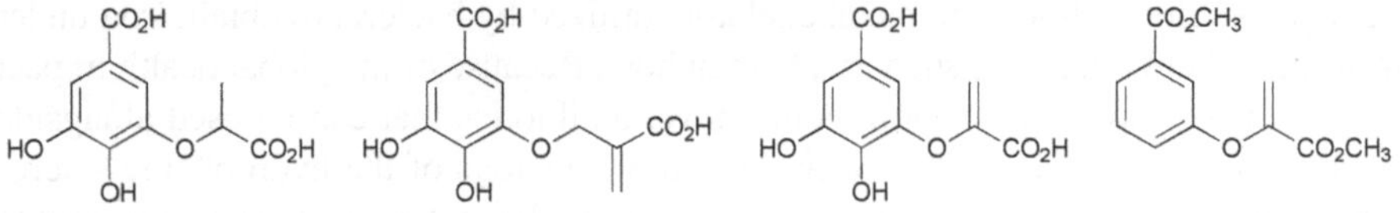

111   **112**   **113**   **114**

Fig. 5.23 Chorismate-based MbtI inhibitors [81]

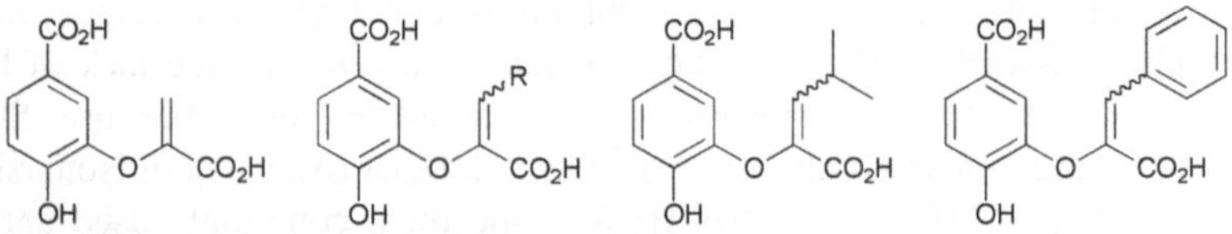

115   **116**   **117**   **118**

Fig. 5.24 Chorismate-based MbtI inhibitors. SAR of the C3 position [81]

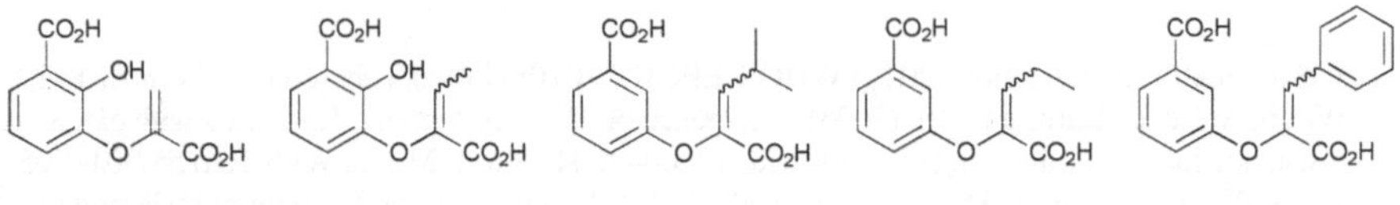

119   **120**   **121**   **122**   **123**

Fig. 5.25 Isochorismate-based MbtI inhibitors. SAR of the C3 position [81]

Substitution at the C3 position was also studied, while preserving the hydroxyl group at C4 in order to determine its importance in biological activity. Analogs **115–118** were prepared as mixtures of inseparable isomers (Fig. 5.24). The isomers were moderate to poor inhibitors of MbtI, with **116** ($K_i = 290 \pm 90 \ \mu M$) and **117** ($K_i = 310 \pm 70 \ \mu M$) displaying the best activity. While the C4 hydroxyl group appeared to have little effect on activity, substitution at C3 improved the activity against MbtI but decreased it in the case of *S. marcescens* AS, providing a potential source of selectivity in inhibitor design.

A series of inhibitors designed to mimic the enzyme-bound isochorismate scaffold was synthesized (Fig. 5.25). As observed before, substitution at the C3

position improved the activity against MbtI with compounds **120–122** displaying potent activity, **120** ($K_i = 11 \pm 1$ μM). These compounds represent the most
potent inhibitors of MbtI or any other salicylate synthase reported to date.

5.10 Conclusions

Iron acquisition has been validated as a biological target to develop novel antibacterial agents. Siderophores are metal chelators utilized by bacteria to obtain iron under
iron-limited environments such as a human host. Because of the global health impact
of TB, there is the need of developing novel antibiotics that can be used alongside
the current first-line drugs. Interestingly, simple analogs of the hydrophobic siderophores used by *M. tuberculosis* have proven to be potent and selective growth
inhibitors, although the specifics regarding their precise mode of action remain to
be elucidated. Exploiting nature's own design, mycobactin-artemisinin conjugate
56 was synthesized and demonstrated to be a potent and highly selective anti-TB
agent (MIC = 0.39 μg/mL, *M. tuberculosis* H$_{37}$Rv). A different approach to target iron acquisition is the synthesis of inhibitors targeting the enzymes involved
in siderophore assembly. The advantage of this work relies on the lack of human
homologues, rendering the analogues as ideal candidates for human use. Specific
examples of these molecules already display activity that rivals that of isoniazid (**69**,
MIC$_{99}$ = 0.19 μM, *M. tuberculosis* H$_{37}$Rv), the most commonly used antibiotic
against *M. tuberculosis* today. Regardless of the approach taken, these molecules
provide the knowledge and tools largely needed to deal with the burden of TB.

References

1. World Health Organization (2010) WHO REPORT 2010: Global Tuberculosis Control 1–205
2. Wright A, Bai G, Barrera L et al (2006) Emergence of *Mycobacterium tuberculosis* with extensive
   resistance to second-line drugs—worldwide (2000–2004). Morb Mortal Wkly Rep 55:301–305
3. Jones PB, Parrish NM, Houston TA et al (2000) A new class of antituberculosis agents. J
   Med Chem 43:3304–3314
4. Lenaerts AJ, Gruppo V, Marietta KS et al (2005) Preclinical testing of the nitroimidazopyran
   PA-824 for activity against *Mycobacterium tuberculosis* in a series of in vitro and in vivo
   models. Antimicrob Agents Chemother 49:2294–2301
5. Nikonenko BV, Protopopova M, Samala R et al (2007) Drug therapy of experimental tuberculosis (TB): improved outcome by combining SQ109, a new diamine antibiotic, with exist-
   ing TB drugs. Antimicrob Agents Chemother 51:1563–1565
6. Parrish NM, Houston T, Jones PB et al (2001) In vitro activity of a novel antimycobacterial compound *n*-octanesulfonylacetamide, and its effects on lipid and mycolic acid synthesis.
   Antimicrob Agents Chemother 45:1143–1150
7. Protopopova M, Hanrahan C, Nikonenko B et al (2005) Identification of a new antitubercular
   drug candidate, SQ109, from a combinatorial library of 1,2-ethylenediamines. J Antimicrob
   Chemother 56:968–974
8. Stover CK, Warrener P, VanDevanter DR et al (2000) A small-molecule nitroimidazopyran
   drug candidate for the treatment of tuberculosis. Nature 405:962–966

9. Tahlan K, Wilson R, Kastrinsky DB et al (2012) SQ109 targets MmpL3, a membrane transporter of trehalose monomycolate involved in mycolic acid donation to the cell wall core of *Mycobacterium tuberculosis*. Antimicrob Agents Chemother 56:1797–1809
10. Van den Boogaard J, Kibiki GS, Kisanga ER et al (2009) New drugs against tuberculosis: problems, progress, and evaluation of agents in clinical development. Antimicrob Agents Chemother 53:849–862
11. Vilchèze C, Baughn AD, Tufariello J et al (2011) Novel inhibitors of InhA efficiently kill *Mycobacterium tuberculosis* under aerobic and anaerobic conditions. Antimicrob Agents Chemother 55:3889–3898
12. Banin E, Lozinski A, Brady KM et al (2008) The potential of desferrioxamine-gallium as an anti-*pseudomonas* therapeutic agent. Proc Natl Acad Sci USA 105:16761–16766
13. Braun V, Pramanik A, Gwinner T et al (2009) Sideromycins: tools and antibiotics. Biometals 22:3–13
14. Chu BC, Garcia-Herrero A, Johanson TH et al (2010) Siderophore uptake in bacteria and the battle for iron with the host; a bird's eye view. Biometals 23:601–611
15. Ji C, Juárez-Hernández RE, Miller MJ (2012) Exploiting bacterial iron acquisition: siderophore conjugates. Future Med Chem 4:297–313
16. Miethke M, Marahiel MA (2007) Siderophore-Based iron acquisition and pathogen control. Mol Biol Rev 71:413–451
17. Kaneko Y, Thoendel M, Olakanmi O et al (2007) The transition metal gallium disrupts *Pseudomonas aeruginosa* iron metabolism and has antimicrobial and antibiofilm activity. J Clin Invest 117:877–888
18. Boukhalfa H, Crumbliss AL (2002) Chemical aspects of siderophore mediated iron transport. Biometals 15:325–339
19. Jurado RL (1997) Infections, and anemia of inflammation. Clin Infect Dis 25:888–895
20. Williams RJP (1990) An introduction to the nature of iron transport and storage. In: Ponka P, Schulman HM, Woodworth RC (eds) Iron transport and storage. CRC Press, Boca Raton
21. Andrews SC, Robinson AK, Rodríguez-Quiñones F (2003) Bacterial iron homeostasis. FEMS Microbiol Rev 27:215–237
22. Banerjee S, Farhana A, Ehtesham NZ et al (2011) Iron acquisition, assimilation and regulation in mycobacteria. Infect Genet Evol 11:825–838
23. Hider RC, Kong XL (2010) Chemistry and biology of siderophores. Nat Prod Rep 27:637–657
24. Raymond KN, Dertz EA (2004) Biochemical and physical properties of siderophores. In: Crosa JH, Mey AR, Payne SM (eds) Iron transport in bacteria. ASM Press, Washington, DC
25. Sandy M, Butler AA (2009) Microbial iron acquisition: marine and terrestrial siderophores. Chem Rev 109:4580–4595
26. Ji C, Miller PA, Miller MM (2010) Iron transport-mediated drug delivery: practical syntheses and in vitro antibacterial studies of tris-catecholate siderophore-aminopenicillin conjugates reveals selectively potent antipseudomonal activity. J Am Chem Soc 134:9898–9901
27. Doorneweerd DD, Henne WA, Reifenberger RG et al (2010) Selective capture and identification of pathogenic bacteria using an immobilized siderophore. Langmuir 26:15424–15429
28. Inomata T, Eguchi H, Matsumoto K (2007) Adsorption of microorganisms onto an artificial siderophore-modified Au substrate. Biosens Bioelectron 22:751–755
29. Snow GA (1965) Isolation and Structure of mycobactin T, a growth factor from *Mycobacterium tuberculosis*. Biochem J 97:166–175
30. Snow GA (1970) Mycobactins: iron-chelating growth factors from mycobacteria. Bacteriol Rev 34:99–125
31. Barry CE III, Boshoff H (2005) Getting the iron out. Nat Chem Biol 1:127–128
32. Luo M, Fadeev EA, Groves JT (2005) Mycobactin-mediated iron acquisition within macrophages. Nat Chem Biol 1:149–153
33. Gobin J, Horwitz MA (1996) Exochelins of *Mycobacterium tuberculosis* remove iron from human iron-binding proteins and donate iron to mycobactins in the *M. tuberculosis* cell wall. J Exp Med 183:1527–1532

34. Gobin J, Moore CH, Reeve JR Jr, Wong DK et al (1995) Iron acquisition by *Mycobacterium tuberculosis*: Isolation and characterization of a family of iron-binding exochelins. Proc Natl Acad Sci USA 92:5189–5193
35. Rodriguez GM (2006) Control of iron metabolism in *Mycobacterium tuberculosis*. Trends Microbiol 14:320–327
36. Quadri LEN (2008) Iron uptake in mycobacteria. In: Daffé M, Reyrat JM (eds) The mycobacterial cell envelope. ASM Press, Washington, DC
37. Rodriguez GM, Smith I (2006) Identification of an ABC transporter required for the iron acquisition and virulence in *Mycobacterium tuberculosis*. J Bacteriol 188:424–430
38. McCready KA, Ratledge C (1979) Ferrimycobactin Reductase Activity from *Mycobacterium smegmatis*. J Gen Microbiol 113:67–72
39. Brown KA, Ratledge C (1975) Iron transport in *Mycobacterium smegmatis*: ferrimycobactin reductase (NAD(P)H:ferrimycobactin oxidoreductase), the enzyme releasing iron from its carrier. FEBS Lett 53:262–266
40. Ratledge C (2004) Iron, mycobacteria and tuberculosis. Tuberculosis 84:110–130
41. Benz G, Schröder T, Kurz J et al (1982) Konstitution der Deferriform der Albomycine δ1, δ2 und ε. Ang Chem Int Ed 94:552–553
42. Clarke TE, Braun V, Winkelmann G et al (2002) X-ray crystallographic structures of the *Escherichia coli* periplasmic protein FhuD bound to hydroxamate-type siderophores and the antibiotic albomycin. J Biol Chem 277:13966–13972
43. Destoumieux-Garzón D, Thomas X, Santamaria M et al (2003) Microcin E492 antibacterial activity: evidence for a TonB-dependent inner membrane permeabilization on *Escherichia coli*. Mol Microbiol 49:1031–1041
44. Duquesne S, Destoumieux-Garzón D, Peduzzi J, Rebuffat S (2007) Microcins, gene-encoded antibacterial peptides from enterobacteria. Nat Prod Rep 24:708–734
45. Lagos R, Wilkens M, Vergara C et al (1993) Microcin E492 forms ion channels in phospholipid bilayer membranes. FEBS Lett 321:145–148
46. Nolan EM, Fischbach MA, Koglin A et al (2007) Biosynthethic tailoring of microcin E492 m: post-translational modification affords an antibacterial siderophore-peptide conjugate. J Am Chem Soc 129:14336–14347
47. Vértesy L, Aretz W, Fehlhaber H-W et al (1995) Antibiotics from *Streptomyces violaceus*, DSM 8286, having a siderophor-aminoglycoside structure. Helv Chim Acta 78:46–60
48. Dong L, Roosenberg JM II, Miller MJ (2002) Total synthesis of desferrisalmycin b. J Am Chem Soc 124:15001–15005
49. Roosenberg JM II, Miller MJ (2000) Total synthesis of the siderophore danoxamine. J Org Chem 65:4833–4838
50. Wencewicz TA (2011) Development of microbe-selective antibacterial agents: from small molecules to siderophores. Ph.D. Dissertation, University of Notre Dame, Notre Dame, IN
51. Wencewicz TA, Möllmann U, Long TE et al (2009) Is drug release necessary for antimicrobial activity of siderophore-drug conjugates? Syntheses and biological studies of the naturally occurring salmycin "Trojan Horse" antibiotics and synthetic desferridanoxamineantibiotic conjugates. Biometals 22:633–648
52. Möllmann U, Dong L, Vértesy L et al (2004) Salmycins-natural siderophore-drug conjugates: prospects for modification and investigation based on successful total synthesis. Paper presented at the 2nd international Biometals symposium, Garmisch-Partenkirchen, Germany
53. Maurer PJ, Miller MJ (1983) Total Synthesis of a mycobactin: mycobactin S2. J Am Chem Soc 105:240–245
54. Hu J, Miller MJ (1997) Total synthesis of mycobactin S, a siderophore and growth promoter of *Mycobacterium smegmatis*, and determination of its growth inhibitory activity against *Mycobacterium tuberculosis*. J Am Chem Soc 119:3462–3468
55. Fennell KA, Möllmann U, Miller MJ (2008) Syntheses and biological activity of amamistatin b and analogs. J Org Chem 73:1018–1024
56. Miller MJ, Walz AJ, Zhu H et al (2011) Design, synthesis, and study of a mycobactin-artemisinin conjugate that has selective and potent activity against tuberculosis and malaria. J Am Chem Soc 133:2076–2079

57. Xu Y, Miller MJ (1998) Total syntheses of mycobactin analogs as potent antimycobacterial agents using a minimal protecting group strategy. J Org Chem 63:4314–4322
58. Zhu H, Miller MJ (Unpublished work, 2013) University of Notre Dame, Notre Dame, IN
59. Walz AJ, Möllmann U, Miller MJ (2007) Synthesis and studies of catechol containing mycobactin S and T analogs. Org Biomol Chem 5:1621–1628
60. Schwynn B, Neilands JB (1987) Universal chemical assay for the detection and determination of siderophores. Anal Biochem 160:47–56
61. Juárez-Hernández RE (2012) Convergent approach for the syntheses of sideromycins: mycobactin T and gallioxamine B conjugates. Ph.D. Dissertation, University of Notre Dame, Notre Dame, IN
62. Juárez-Hernández RE, Franzblau SG, Miller MJ (2012) Syntheses of mycobactin analogs as potent and selective inhibitors of *Mycobacterium tuberculosis*. Org Biomol Chem 10:7584–7593
63. Moraski GC, Markley LD, Chang M et al (2012) Generation and exploration of new classes of antitubercular agents: the optimization of oxazolines, oxazoles, thiazolines, thiazoles to imidazo [1,2-*a*] pyridines and isomeric 5,6-fused scaffolds. Bioorg Med Chem Lett 20:2214–2220
64. Moraski GC, Chang M, Villegas-Estrada A et al (2010) Structure-activity relationship of new antituberculosis agents derived from oxazoline and oxazole esters. Eur J Med Chem 45:1703–1716
65. Moraski GC, Franzblau SG, Miller MJ (2010) Utilization of the suzuki coupling to enhance the antituberculosis activity of aryloxazoles. Heterocycles 80:977–988
66. Moraski GC, Markley LD, Hipskind PA et al (2011) Advent of Imidazo[1,2-a]pyridine-3-carboxamides with Potent Multi- and Extended Drug Resistant Antituberculosis Activity. ACS Med Chem Lett 2:466–470
67. De Voss JJ, Rutter K, Schroeder BG et al (1999) Iron acquisition and metabolism by mycobacteria. J Bacteriol 181:4443–4451
68. De Voss JJ, Rutter K, Schroeder BG et al (2000) The salicylate-derived mycobactin siderophores of *Mycobacterium tuberculosis* are essential for growth in macrophages. Proc Natl Acad Sci USA 97:1252–1257
69. McMahon MD, Rush JS, Thomas MG (2012) Analyses of mycobactin TB, mycobactin TE, and mycobactin TF suggest revisions to the mycobactin biosynthesis pathway in *Mycobacterium tuberculosis*. J Bacteriol 194:2809–2818
70. Quadri LEN, Sello J, Keating TA et al (1998) Identification of a *Mycobacterium tuberculosis* gene cluster encoding the biosynthetic enzymes for assembly of the virulence-conferring siderophore mycobactin. Chem Biol 5:631–645
71. Sieber SA, Marahiel MA (2005) Molecular mechanisms underlying nonribosomal peptide synthesis: approaches to new antibiotics. Chem Rev 105:715–738
72. Gulick AM, Lu X, Dunaway-Mariano D (2004) Crystal structure of 4-chlorobenzoate: CoA ligase/synthetase in the unliganded and aryl substrate-bound states. Biochemistry 43:8670–8679
73. Somu RV, Boshoff H, Qiao C et al (2006) Rationally designed nucleoside antibiotics that inhibit siderophore biosynthesis of *Mycobacterium tuberculosis*. J Med Chem 49:31–34
74. Finking R, Neumüller A, Solsbacher J et al (2003) Aminoacyl adenylate substrate analogues for the inhibition of adenylation domains of nonribosomal peptide synthetases. Chem Bio Chem 4:903–906
75. May JJ, Finking R, Wiegeshoff F et al (2005) Inhibition of the D-alanine: D-alanyl carrier protein ligase from *Bacillus subtilis* increases the bacterium's susceptibility to antibiotics that target the cell wall. FEBS J 272:2993–3003
76. Ferreras JA, Ryu J-S, Di Lello F et al (2005) Small-molecule inhibition of siderophore biosynthesis in *Mycobacterium tuberculosis* and *Yersinia pestis*. Nat Chem Biol 1:29–32
77. Vannada J, Bennett EM, Wilson DJ et al (2006) Design, synthesis, and biological evaluation of β-ketosulfonamide adenylation inhibitors as potential antitubercular agents. Org Lett 8:4707–4710
78. Somu RV, Wilson DJ, Bennett EM et al (2006) Antitubercular nucleosides that inhibit siderophore biosynthesis: SAR of the glycosyl domain. J Med Chem 49:7623–7635

79. Neres J, Labello NP, Somu RV et al (2008) Inhibition of siderophore biosynthesis in *Mycobacterium tuberculosis* with nucleoside bisubstrate analogues: structure-activity relationships of the nucleobase domain of 5'-*o*-[*n*-(salicyl)sulfamoyl] adenosine. J Med Chem 51:5349–5370
80. Harrison AJ, Yu M, Gårdenborg T et al (2006) The structure of mycobactin TI from *Mycobacterium tuberculosis*, the first enzyme in the biosynthesis of the siderophore mycobactin, reveals it to be a salicylate synthase. J Bacteriol 188:6081–6091
81. Manos-Turvey A, Bulloch EMM, Rutledge PJ et al (2010) Inhibition Studies of *Mycobacterium tuberculosis* Salicylate Synthase (MbtI). ChemMedChem 5:1067–1079

Index

B. R. Byers (ed.), *Iron Acquisition by the Genus Mycobacterium*,
SpringerBriefs in Biometals, DOI: 10.1007/978-3-319-00303-0,
© The Author(s) 2013